STUDIES IN OPTIMIZATION

Studies in Mathematics

The Mathematical Association of America

R. W. Cottle
Stanford University

G. B. Dantzig
Stanford University

B. C. Eaves
Stanford University

Jack Edmonds
University of Waterloo

D. R. Fulkerson
Cornell University

H. W. Kuhn
Princeton University

Herbert Scarf
Yale University

Lloyd Shapley
Rand Corporation, Santa Monica

A. W. Tucker
Princeton University

A. F. Veinott, Jr.
Stanford University

Philip Wolfe
IBM, Yorktown Heights

Studies in Mathematics

Volume 10

STUDIES IN OPTIMIZATION

G. B. Dantzig and
B. C. Eaves, editors

Stanford University

Published and distributed by

The Mathematical Association of America

CONTENTS

ACKNOWLEDGMENTS

Professor Tucker's contribution, "Combinatorial Algebra of Linear Programs," is a slight revision of a paper, prepared with the assistance of M. Balinski, H. D. Mills, and R. R. Singleton, and published in NEW DIRECTIONS IN MATHEMATICS (Dartmouth College Mathematics Conference, 1961), edited by J. G. Kemeny, R. Robinson, and R. W. Ritchie, Prentice-Hall 1963, pp. 77–91; Professor Cottle's and Professor Dantzig's contribution, "Complementary Pivot Theory of Mathematical Programming," previously appeared in MATHEMATICS OF THE DECISION SCIENCES, Part 1, Lectures in Applied Mathematics, Volume 11, published by AMS, 1968, and appeared also in LINEAR ALGEBRA AND ITS APPLICATIONS, Volume 1, 1968; Professor Edmonds' and Professor Fulkerson's contribution, "Bottleneck Extrema," previously appeared in the JOURNAL OF COMBINATORIAL THEORY, Volume 8, Number 3, 1970; Professor Scarf's and Dr. Shapley's contribution, "On Cores and Indivisibility," previously appeared in the JOURNAL OF MATHEMATICAL ECONOMICS, Volume 1, 1974; Professor Dantzig's and Dr. Wolfe's contribution, "The Decomposition Algorithm for Linear Programs," previously appeared in ECONOMETRICA, Volume 29, Number 4, 1961.

STUDIES IN OPTIMIZATION

INTRODUCTION

This volume of the Mathematical Association of America series Studies in Mathematics is a collection of eight papers from the field of mathematical optimization. This collection constitutes an interesting sample of topics and techniques of current interest in optimization, however, we caution, it is too small to be considered representative of this important and developing subject.

As practiced, mathematical optimization can be approximately described as the tasks of

(1) Developing a mathematical structure, called a program, which models some "real world situation"; in general, relations of the structure represent restrictions on the value of the variables and the objective function(s) provides a measure(s) of performance.

(2) Investigating existence and attributes of optimal (or near optimal) solutions; finding ways to characterize optimal policies.

(3) Designing and utilizing algorithms for computation of optimal (or near optimal) solutions.

(4) Implementing the mathematical solution in a particular application, evaluating the results, and making modifications.

Each paper of this volume treats some aspects of tasks (1), (2) and (3). After making brief comments about each paper, we will say a few words about the relationship of optimization theory to an education in mathematics.

The first paper by A. W. Tucker, "Combinatorial Algebra of Linear Programs" opens and closes with discussions regarding the growing impact of combinatorics in applied areas of mathematics. As linear programming is perhaps the main prerequisite for appreciating the other articles in this volume, a reader might well begin his perusing with this article.

One device for studying a mathematical program is through duality theory. Given an optimization problem, called the *primal* program, there is often another optimization problem, called the *dual*, which is related in a very strong and intricate way to the original problem; that is, to the primal program. For example, the optimal objective value of the primal and dual programs might be equal or a solution to the dual might yield a solution to the primal. Most of the papers in this volume, e.g., that of Tucker, make extensive use of duality concepts.

The second paper by Richard W. Cottle and George B. Dantzig, "Complementary Pivot Theory of Mathematical Programming", defines the linear complementarity problem as that of seeking n-vectors w and z satisfying:

$$w = q + Mz \qquad w \geqq 0 \qquad z \geqq 0$$
$$w \cdot z = 0$$

where M, a square matrix, and q, a vector, are given. This model is shown to include linear and quadratic programming and general sum two-person games in normal form. Lemke's finite complementary pivoting algorithm for finding a complementary solution for certain M and q is next described and established. As the simplex method for linear programming the scheme iterates by moving from one extreme point to the next along edges of a polyhedral set. However, Lemke's algorithm is most unusual in that convergence is not based on monotone improvement of some function. It is this feature that permits one to find globally optimal solutions to certain classes of non-linear non-convex extremum problems.

Harold W. Kuhn, in his article, " 'Steiner's' Problem Revisited", treats the problem of finding a point, popularly called a Steiner Point, in Euclidean space such that the weighted sum of distances from this point to n given points is minimum. The Steiner Point is shown to have certain attributes. The non-linear dual problem is developed and shown to have an interesting form. An algorithm for solving the problem is described and its convergence to the unique solution is proved.

The article by B. Curtis Eaves, "Properly Labeled Simplexes", treats a generalization of Sperner's Lemma; triangulations are needed in the proof but not in the statement of this result. The unique convergence proof of Lemke for complementary pivoting, as discussed in the article of this volume by Cottle and Dantzig, is employed on an infinite structure. An algorithm for computing fixed points is a natural by-product of the method of proof.

Jack Edmonds and D. R. Fulkerson, in their article, "Bottleneck Extrema", develop a "min max" problem

$$\min_{R \in \mathcal{R}} \max_{x \in R} f(x)$$

wherein they minimize over certain subsets R of a finite set E and maximize over elements x of the subsets R. For example, one might consider the task of finding a route through a network between two given points such that the bottleneck (that is, the least arc number encountered on the route) is least restrictive. A dual "max min" problem

$$\max_{S \in \mathcal{S}} \min_{x \in S} f(x)$$

is treated wherein they maximize over certain subsets of the same finite set E and minimize over elements of the subset. The main results are a theorem that the primal and dual problems have equal objective value and an algorithm for computing solutions to these two problems. Note that the duality theorem here, for example, is based on a discrete or combinatorial structure whereas others, as in Kuhn's paper, are based on continuous or nondiscrete structures.

Herbert Scarf and Lloyd S. Shapley, in their article, "On Cores and Indivisibility", develop a model for a market in which n participants trade indivisible goods; each participant (a trader) enters the market with one unit of a good, say a house, and his preference ordering of all goods of the market. It is proved that the "core" is nonempty, that is, it is proved that there is a scheme of trading among the n participants such that no subcollection of the participants, in view of their preferences, might wish to withdraw to trade only among themselves. The development is based upon a theorem of Scarf; a proof of this theorem is given in Example 4 of "Properly Labeled Simplexes" in this volume.

The article by Arthur F. Veinott, Jr., "Markov Decision Chains", treats the problem of optimally controlling a finite state Markov chain process. If the process is initiated in a given state, then one of a finite number of alternatives must be selected which results in an "immediate income" (or expense) but also determines the probability distribution of the transition to the next state. An infinite sequence of immediate incomes is thus generated. The paper concerns itself with evaluating these income sequences, under various criteria, investigating the properties of optimal solutions, and computing optimal solutions. Applications include a gambling problem, a sequential decision problem, and an inventory problem.

The last paper, "The Decomposition Algorithm for Linear Programs", by George B. Dantzig and Philip Wolfe, provides a device whereby one large linear program is partitioned into two sets, one representing a set of "joint" constraints and the second set consisting of a collection of smaller subproblems that are independent of each other except that their variables are related through the joint constraints. Following partition one has associated with the joint constraints a "master" program which coordinates the solutions of a collection of subproblem programs. The coordination between the master and subproblems proceeds roughly as follows: the master delivers to a subproblem tentative "prices" (Lagrange Multipliers) associated with the joint constraints; assuming these prices, the subproblem optimizes and returns to the master problem certain facts about this optimal solution. Using this new information the master linear program is resolved. The process repeats and terminates with an optimal solution to the original large linear program. This decomposition principle "yields a certain rationale for the decentralized decision process in the theory of the firm."

The remainder of this introduction is devoted to a recommendation regarding the future of optimization theory in mathematics education.

Optimization as we have described it became viable with the advent of computers. The computer revolution has so broadened the base of quantitative analysis that almost all areas of human endeavor are now being modelled in mathematical terms. It is said that approximately one-fourth of all current scientific computation involves optimization. It is this force that has spurred the rapid progress in this field.

Optimization theory is now a fertile ground for new and pressing problems, for classes of problems upon which to build new mathematical theories; it could be a source of new, exciting, and relevant problems which would serve to stimulate and to motivate the mathematics student. The creative student could be challenged by the collection of outstanding unsolved problems. Two such problems are the traveling salesman problem and the Hirsh (or m-step) conjecture; both of these problems are concerned with finding a "good algorithm" in the sense of J. Edmonds: a good algorithm is one in which the time to compute a solution to a numerical problem grows algebraically (e.g., not exponentially) with the size of the problem.

The shortest route problem is related to the traveling salesman problem but is much easier: given a road map, consider the task of finding the shortest distance from San Francisco to Boston. One formulates this problem in mathematical terms by representing the cities as points in a set and roads between cities by arcs, a binary relation. A distance is assigned to the arcs and the concept of a simple path in a graph is introduced. Since the number of simple paths between any two points is obviously finite, the shortest route problem is uninteresting from an existence point of view. Namely, just pick the shortest path among the finite set of possible paths.

Unfortunately for the case of a complete graph with n nodes there are many such paths. Using direct enumeration for $n = 90$, if there were 10^{30} computers on each of the 10^{11} stars of the Milky Way, if each computer analyzed 10^{30} paths per second, and if all these computers were running in parallel since the beginning of time 10^{10} years ago (the big bang), only a tiny fraction of the paths would, to date, have been examined. Given m nodes and n arcs, can one devise a shortest route algorithm that requires at most, say $n + m$ additions and comparisons; what is the best one can do? It is known that $n^2/2$ additions and n^2 comparisons can be attained for the complete graph with non-negative distances. Thus for the shortest route problem good algorithms have been devised provided the arc distances are non-negative. If the arc-distances can have negative as well as positive values does there exist a "good" shortest path algorithm? The enlarged class of problems now includes the famous traveling salesman problem: a traveling salesman has a sweetheart in the capital city of each of the 50 states. He feels it is his duty to arrange his tour so as to visit each of his sweethearts. Find the shortest tour. This problem has been lying about in mathematical circles since the 1930's and perhaps earlier. From an algorithmic point of view this is a fascinating problem.

It is now known that the simplex method of linear programming is not a "good algorithm" in the sense of Edmonds. Nevertheless, for tens of thousands of applied problems the simplex method has converged in the neighborhood of $3m$ steps where m is the number of equations in the linear program. However, examples have been constructed where the number of iterations grows exponentially with n, the number of variables. A full explanation of the difference between practiced and contrived examples appears to be beyond the capabilities of current mathematics. An interesting problem related to this general question is the Hirsh (or m-step) conjecture. Namely, given two extreme points p and q of an n-dimensional polyhedral convex set P with f full dimensional faces; does there exist a path of extreme edges from p to q with no more than $m = f - n$ edges? This existence question is easy enough to understand, as is the problem of the traveling salesman, and as elusive.

It seems to us that omission of optimization in the typical mathematics curriculum deprives the student of an exciting and relevant outlet for their talents. Recently a combinatorial course at a major university was informally surveyed to find out how many had even heard of such subjects as linear programming, mathematical programming, network theory, or integer programming. Only about half the class had. Those who had not were mathematics majors.

In order to expose the student to mathematical optimization we have two suggestions. *First*, new courses could supplement (or replace) those in the traditional undergraduate curriculum in mathematics; for example, an introductory course on finite mathematics (already popular in many schools) or an advanced undergraduate course on combinatorial analysis which includes some optimization theory. Classical combinatorics is sometimes useful—it occasionally prevents people from programming an exhaustive search procedure on the computer. For example, one would probably decide not to list out all the paths in a network, if he knew in advance that there were an astronomical number of possibilities. However, the portion of combinatorial analysis which seems to have the most important applications is that concerned with selecting the best of all the combinations. This, for example, is what linear programming is all about. An economy has many alternative technologies it can draw upon; some use more labor than others and make more intense use of scarce resources, or more intensive use of limited capacity. The problem then becomes how to select, how much, and when.

Second, aspects of university organization might be revised so that all students of mathematics are more likely to be exposed, as a matter of curricular routine, to the "mathematical sciences". At present, there is no general descriptive term to cover the fields of operations research, management science, control theory, statistics, numerical analysis as applied to computer science, and traditional applied mathematics. What is emerging instead is one sweeping descriptive term "Mathematical Sciences" which also includes "pure" mathematics. Universities might consider developing a

School of Mathematical Sciences encompassing mathematics, operations research, statistics, computer science, and classical applied mathematics.

As an historical note, a number of fields would be broadly referred to as belonging to "Applied Mathematics" except for a semantic difficulty: the term "Applied Mathematics" has been used, traditionally, for mathematical systems drawn from the physical sciences. New terminology has emerged to circumvent this semantic impasse, for example "Mathematical Sciences" and "Operations Research."

GEORGE B. DANTZIG
B. CURTIS EAVES

COMBINATORIAL ALGEBRA OF LINEAR PROGRAMS*

A. W. Tucker

Many advances on the frontiers of mathematics are related increasingly to the "applied" interests of social scientists, economists, statisticians, operations researchers, industrial and design engineers, and others. Such subjects as linear programming, game theory, network theory, Boolean algebra, Markov processes, and information theory are used frequently in industrial and government applications, and appear increasingly often both as topics for research and as tools in investigations in other areas. These and other related subjects fall mainly within the rapidly developing field of Combinatorial Mathematics. The author's view of the importance of this field was given in the Foreword to the Novem-

* This is a slight revision of a paper, prepared with the assistance of M. L. Balinski, H. D. Mills, and R. R. Singleton, and published in NEW DIRECTIONS IN MATHEMATICS (Dartmouth College Mathematics Conference, 1961), edited by J. G. Kemeny, R. Robinson and R. W. Ritchie, Prentice-Hall 1963, pp. 77–91.

ber, 1960, issue of the *IBM Journal,* which was especially devoted
to the field.

"Combinatorial Mathematics, or 'Combinatorics', regarded as
originating in the *Ars Combinatoria* of Leibniz, has to do with
problems of arrangement, operation, and selection within a finite
or discrete system—such as the aggregate of all possible states of a
digital computer. Until recently, preoccupation with continuous
mathematics has inhibited the growth of discrete mathematics.
But now it is realized that combinatorial methods can be developed
to attack profitably, in modern science and technology, a vast
variety of 'problems of organized complexity'—an apt designation
of Warren Weaver [1]. In 1947 Hermann Weyl [2] wrote as follows (rearranged slightly for quotation here):

" 'Perhaps the philosophically most relevant feature of modern
science is the emergence of abstract symbolic structures as the hard
core of objectivity behind—as Eddington puts it—the colorful
tale of the subjective storyteller mind. The combinatorics of ag-
gregates and complexes deals with some of the simplest such structures imaginable. It is gratifying that combinatorial mathematics
is so closely related to the philosophically important problems of
individuation and probability, and that it accounts for some of the
most fundamental phenomena in inorganic and organic nature.
This structural viewpoint occurs in the foundations of quantum
mechanics. In a widely different field John von Neumann's and
Oskar Morgenstern's attempt to found economics on a theory of
games is characteristic of the same trend. The network of nerves
joining the brain with the sense organs is a subject that by its very
nature invites combinatorial investigation. Modern computing
machines translate our insight into the combinatorial structure of
mathematics into practice by mechanical and electronic devices.' "

Let us now examine one specific new direction in applied mathematics which exhibits combinatorial structure characteristic of the
field outlined above. This is *linear programming,* a subject born in
1947 and now extended in various ways under the title "mathematical programming"—to avoid confusion with computer programming. This subject has an extensive literature from which

four items are selected rather arbitrarily for mention:—a complete treatise [3], with introductory chapters on problems, origins, and models, by the originator of the subject, and three elementary textbooks, [4], [5], and [6], with linear-programming chapters using the same tableau-pivot format we do here.

A MINIATURE EXAMPLE OF LINEAR PROGRAMMING

A *linear program* is a problem of "optimizing" (i.e., maximizing or minimizing) a linear "objective" function of many (real) variables subject to a system of linear "constraints," each of which is a linear inequality or linear equation. The following is an example, in miniature,

Minimize the objective function

$$(0) \quad -\lambda + 3\mu$$

subject to the constraints

$$(1) \quad -\lambda + 2\mu \geqq 2$$
$$(2) \quad \lambda - \mu \geqq -3$$
$$(3) \quad \mu \geqq 1$$
$$(4) \quad \lambda - 2\mu \geqq -5$$
$$(5) \quad -\lambda + \mu \geqq 2.$$

This miniature linear program can readily be analyzed graphically. In a λ, μ-coordinate plane (see Figure 1) we plot the "halfplanes" (1)–(5). In each case the halfplane (which includes its boundary line) is labelled by its number along its boundary line on the side of the line in which the halfplane lies: for example, the label (1) appears along the line $-\lambda + 2\mu = 2$ within the halfplane $-\lambda + 2\mu \geqq 2$. The set of *feasible* points (λ, μ) satisfying all five constraints is a quadrilateral determined by (2), (3), (4), (5). The first constraint is inactive, as it happens, because the feasible set lies entirely within the "open halfplane" $-\lambda + 2\mu > 2$. The

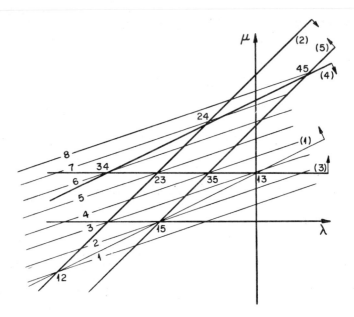

FIG. 1. Graphical analysis of the miniature linear program.

objective function, which is to be minimized, is represented by some particular contour (or level) lines along which it takes values as indicated. We see that the desired minimum appears to occur at the point of intersection of lines (3) and (5), namely at the point 35, with coordinates $\lambda = -1$, $\mu = 1$, where the objective function takes the value 4.

SYSTEMATIC DISCUSSION OF EXAMPLE

The foregoing graphical method of solving a linear program does not generalize beyond two or three variables, whereas linear programs involving scores or hundreds of variables are not uncommon in actual practice. So, starting afresh, we outline a tabular systematization of our example which does generalize to large-scale problems.

To this end, we introduce the "tableau"

	y_1	y_2	y_3	y_4	y_5	-1	
λ	-1	1	0	1	-1	-1	$=0$
μ	2	-1	1	-2	1	3	$=0$
-1	2	-3	1	-5	2	0	$=v$
	$=x_1$	$=x_2$	$=x_3$	$=x_4$	$=x_5$	$=u$	

consisting of a *matrix* of three rows and six columns inside the box and certain *marks* λ, μ, -1, etc., around the four margins of the box. The six signs of equality along the bottom of the box indicate a "column system" of six linear equations

$$-\lambda + 2\mu - 2 = x_1$$

$$\lambda - \mu + 3 = x_2$$

$$\mu - 1 = x_3$$

$$\lambda - 2\mu + 5 = x_4$$

$$-\lambda + \mu - 2 = x_5$$

$$-\lambda + 3\mu = u$$

obtained by forming inner (or scalar) products of λ, μ, -1 with the columns of the matrix and setting these inner products equal to x_1, x_2, x_3, x_4, x_5 and u. Then the constraints (1)–(5) of our example, as previously stated, become $x_1 \geqq 0$, $x_2 \geqq 0$, $x_3 \geqq 0$, $x_4 \geqq 0$, $x_5 \geqq 0$, and the objective function becomes u. We seek values of λ and μ that minimize u and make the five x's nonnegative.

The tableau contains also a "row system" of three linear

equations

$$-y_1 + y_2 \qquad + y_4 - y_5 + 1 = 0$$
$$2y_1 - y_2 + y_3 - 2y_4 + y_5 - 3 = 0$$
$$2y_1 - 3y_2 + y_3 - 5y_4 + 2y_5 \qquad = v,$$

indicated by the three signs of equality at the right-hand margin
of the box. We obtain these equations by forming inner products
of the rows of the matrix with y_1, y_2, y_3, y_4, y_5, -1 and setting these
inner products equal to 0, 0, and v. This row system is essential to
our analysis; it will give rise to a second linear program, "dual"
to our first.

We observe, by substituting for the x's and u from the column
system and then reducing via the row system, that

$$x_1 y_1 + x_2 y_2 + x_3 y_3 + x_4 y_4 + x_5 y_5 - u$$
$$= \lambda(-y_1 + y_2 \qquad + y_4 - y_5 + 1)$$
$$+ \mu(2y_1 - y_2 + y_3 - 2y_4 + y_5 - 3)$$
$$- (2y_1 - 3y_2 + y_3 - 5y_4 + 2y_5 \qquad)$$
$$= \lambda(0) + \mu(0) - (v).$$

Thus, the inner product of the marks at the bottom margin of the
tableau with the marks at the top margin is equal to the inner
product of the marks at the left margin with the marks at the right
margin. By rearranging terms, we have

$$u - v = x_1 y_1 + x_2 y_2 + x_3 y_3 + x_4 y_4 + x_5 y_5$$

for any solutions of the column and row systems. This fundamental equation, in which λ and μ do not appear explicitly, we
call the "key equation" (of duality).

A solution $(\lambda, \mu; x_1, x_2, x_3, x_4, x_5, u)$ of the column system, or
"column-solution," is *feasible* if its five x's are nonnegative. Similarly, a solution $(y_1, y_2, y_3, y_4, y_5, v)$ of the row system, or "row-
solution," is *feasible* if its five y's are nonnegative. We see from the
key equation that

$$u \geqq v$$

for a feasible column-solution and a feasible row-solution. There-fore, the v from some feasible row-solution is a lower bound on the u's from all feasible column-solutions. Moreover, if we can find a feasible column-solution and a feasible row-solution such that

$$u = v,$$

then we have a pair λ, μ which minimizes $u = -\lambda + 3\mu$ subject to the five constraints (1)–(5).

We have now shown that a *sufficient* condition that a feasible row-solution minimize u is that there exist a feasible column-solution such that $u = v$. It can be shown (but not here) that this is also a *necessary* condition. Turning to v as objective function, we see that the values of v for all feasible row-solutions are bounded above by the value of u from any feasible column-solution. Hence a feasible row-solution maximizes v if there is a feasible column-solution such that $u = v$. Thus, the problem of maximizing v over the set of feasible row-solutions is "dual" to the problem of mini-mizing u over the set of feasible column-solutions.

Motivated by the above discussion, we examine the purely alge-braic problem of finding a feasible column-solution and a feasible row-solution such that

$$u - v = x_1y_1 + x_2y_2 + x_3y_3 + x_4y_4 + x_5y_5 = 0.$$

The sum of the nonnegative terms x_iy_i is zero if, and only if, each term $x_iy_i = 0$. But $x_iy_i = 0$ if, and only if, $x_i = 0$ or $y_i = 0$ (or both). So we seek solutions of the column and row systems such that certain of the x's are zero, the complementary y's are zero, and all the remaining x's and y's are nonnegative. The column system has *two* "degrees of freedom" since it consists of six equa-tions in eight unknowns, and the row system has *three* "degrees of freedom" since it consists of three equations in six unknowns. Accordingly, we plan to set *two* of the x's equal to zero and solve for the three remaining x's, and set the complementary three of the y's equal to zero and solve for the two remaining y's.

The number of possible pairs of x's (and complementary triples of y's) is $5!/2! \, 3! = 10$. In our example two of the ten possibilities fail, namely $x_1 = 0$, $x_4 = 0$ and $x_2 = 0$, $x_5 = 0$, because the resulting equations are inconsistent. [This corresponds to the graphical fact that lines (1) and (4) are parallel and lines (2) and (5) are parallel in Figure 1.] The eight possible cases turn out to be as listed in Table 1 (where, for example, case 24 corresponds to taking $x_2 = 0$ and $x_4 = 0$).

Case 35, where $x_3 = 0$, $x_5 = 0$ and $y_1 = 0$, $y_2 = 0$, $y_4 = 0$, yields *nonnegative* values for the remaining x's and y's, namely, $x_1 = 1$, $x_2 = 1$, $x_4 = 2$ and $y_3 = 2$, $y_5 = 1$. These x- and y-quintuples are both feasible (the x's resulting from $\lambda = -1$, $\mu = 1$) and so yield a solution to our example, with $u = 4$ as the desired minimum (and $v = 4$ as the maximum for the dual problem). Notice that the eight cases in Table 1 correspond to the eight points of intersection 12, 13, 15, 23, 24, 34, 35, 45 in Figure 1.

Thus our problem, which may have seemed initially to be an analytic one (of "calculus type") has been solved by a systematic finite search, essentially combinatorial in nature.

CONDENSED TABLEAUS

We now develop an organized procedure for obtaining the information summarized in Table 1. We transform from the initial tableau to a class of condensed tableaus, each of which exhibits column and row systems having the same (x, u)- and (y, v)-solutions as the column and row systems of the initial tableau. Instead of equating to zero a pair of x's and then solving for the remaining x's and u, we will solve the corresponding pair of column equations (in the initial tableau) for λ, μ and use these solutions to eliminate λ, μ from the remaining column equations. Similarly, instead of equating to zero the complementary triple of y's and then solving for the remaining y's and v, we will solve the first two row equations (in the initial tableau) for the remaining two y's and use these solutions to eliminate these two y's from the third row equation.

TABLE 1

Case	λ,	μ	x_1,	x_2,	x_3,	x_4,	x_5	u	x-feasible?	y_1,	y_2,	y_3,	y_4,	y_5	v	y-feasible?
12	-4,	-1	0,	0,	-2,	3,	1	1	no	2,	1,	0,	0,	0	1	yes
13	0,	1	0,	2,	0,	3,	-1	3	no	1,	0,	1,	0,	0	3	yes
15	-2,	0	0,	1,	-1,	3,	0	2	no	2,	0,	0,	0,	-1	2	no
23	-2,	1	2,	0,	0,	1,	1	5	yes	0,	-1,	2,	0,	0	5	no
24	-1,	2	3,	0,	1,	0,	1	7	yes	0,	1,	0,	-2,	0	7	no
34	-3,	1	3,	-1,	0,	0,	2	6	no	0,	0,	1,	-1,	0	6	no
35	-1,	1	-1,	1,	0,	2,	0	4	yes	0,	0,	2,	0,	1	4	yes
45	1,	3	3,	1,	2,	0,	0	8	yes	0,	0,	0,	-2,	-1	8	no

For example, from the elimination process indicated by

	y_1	y_2	y_3	y_4	y_5	-1	
λ	-1	1	0	1	-1	-1	$=0$
μ	2	-1	1	-2	1	3	$=0$
-1	2	-3	1	-5	2	0	$=v$
	$=x_1$	$=x_2$	$=x_3$	$=x_4$	$=x_5$	$=u$	

$\left.\begin{array}{}=0\\=0\end{array}\right\}$I (Solve I for y_1, y_2)

$=v\}$II (Then substitute in II to eliminate y_1, y_2)

      I   II

(Solve I for   (Then substitute in II to
λ, μ)   eliminate λ, μ)

we obtain equivalent systems of equations exhibited by

	0	0	y_3	y_4	y_5	-1	
x_1	1	1	1	-1	0	2	$=-y_1$
x_2	2	1	1	0	-1	1	$=-y_2$
-1	4	1	2	-3	-1	-1	$=v$
	$=\lambda$	$=\mu$	$=x_3$	$=x_4$	$=x_5$	$=u$	

Now, by setting $x_1 = 0$ and $x_2 = 0$, we read off

$$\lambda = -4,\ \mu = -1,\ x_3 = -2,\ x_4 = 3,\ x_5 = 1,\ u = 1,$$

and by setting $y_3 = 0$, $y_4 = 0$, $y_5 = 0$, we read off

$$y_1 = 2,\ y_2 = 1,\ v = 1.$$

Here we have y-feasibility but not x-feasibility. (Compare case 12, Table 1.)

The first two columns of the new tableau serve only to state the two equations which express λ and μ in terms of x_1, x_2. Deleting these, we get a "condensed" (or reduced) tableau

(12)

	y_3	y_4	y_5	-1	
x_1	1	-1	0	2	$=-y_1$
x_2	1	0	-1	1	$=-y_2$
-1	2	-3	-1	-1	$=v$
	$=x_3$	$=x_4$	$=x_5$	$=u$	

that exhibits all the information pertinent to the key equation (in which λ and μ do not appear).

If we had eliminated λ and μ via x_2 and x_3 instead of via x_1 and x_2, and had solved the first two equations for y_2 and y_3 instead of y_1 and y_2, we would have obtained the (full) tableau

	0	0	y_1	y_4	y_5	-1	
x_2	1	0	-1	1	-1	-1	$=-y_2$
x_3	1	1	1	-1	0	2	$=-y_3$
-1	2	-1	-2	-1	-1	-5	$=v$
	$=\lambda$	$=\mu$	$=x_1$	$=x_4$	$=x_5$	$=u$	

and the corresponding condensed tableau

$$
\begin{array}{c}
(23) \qquad y_1 \qquad y_4 \qquad y_5 \qquad -1 \\[4pt]
\begin{array}{c|ccc|c|l}
x_2 & -1 & 1 & -1 & -1 & = -y_2 \\
x_3 & 1 & -1 & 0 & 2 & = -y_3 \\
\hline
-1 & -2 & -1 & -1 & -5 & = v \\
\end{array} \\[6pt]
\quad = x_1 \quad = x_4 \quad = x_5 \quad = u
\end{array}
$$

Setting $x_2 = 0$, $x_3 = 0$ and $y_1 = 0$, $y_4 = 0$, $y_5 = 0$ we read off

$$\lambda = -2, \mu = 1$$

from the full tableau, and we read off

$$x_1 = 2, x_4 = 1, x_5 = 1, u = 5 \quad \text{and} \quad y_2 = -1, y_3 = 2, v = 5$$

from either the full or the condensed tableau. Here we have x-feasibility but not y-feasibility. (Compare case 23, Table 1.)

It is possible, however, to pass directly from (12) to (23). From the process indicated by

$$
\begin{array}{c}
(12) \qquad y_3 \qquad y_4 \qquad y_5 \qquad -1 \\[4pt]
\begin{array}{c|ccc|c|l}
x_1 & 1 & -1 & 0 & 2 & = -y_1 \\
x_2 & 1 & 0 & -1 & 1 & = -y_2 \\
\hline
-1 & 2 & -3 & -1 & -1 & = v \\
\end{array} \\[6pt]
\quad = x_3 \quad = x_4 \quad = x_5 \quad = u
\end{array}
$$

Solve the y_1-equation for y_3, and substitute in the two remaining equations to eliminate y_3

Solve the x_3-equation for x_1, and substitute in the three remaining equations to eliminate x_1

we obtain the equivalent systems of equations summarized by

(23)

	y_1	y_4	y_5	-1	
x_3	1	-1	0	2	$=-y_3$
x_2	-1	1	-1	-1	$=-y_2$
-1	-2	-1	-1	-5	$=v$
	$=x_1$	$=x_4$	$=x_5$	$=u$	

In the transition from (12) to (23) the marginal marks x_1, x_3 and $-y_1$, y_3 are replaced by x_3, x_1 and $-y_3$, y_1, and all remaining marginal marks are unchanged, while the interior entries are transformed appropriately. This is a particular instance of an important but simple operation called a "pivot step," which can be described more generally as follows:

Here the marginal marks x^*, ξ^* and y^*, η^* are replaced by ξ^*, x^* and $-\eta^*$, $-y^*$ and all other marks (such as x, ξ and y, η) are unchanged. This interchange of starred marginal marks only signalizes the fact that the transition from the general tableau on the left to that on the right is accomplished by solving the ξ^*-equation (column) and the y^*-equation (row) on the left for x^* and η^*, requiring $a \neq 0$, and then substituting for x^* and η^* in all the remaining column and row equations to eliminate x^* and η^* from them. The formal rules for the transformation of interior entries in the above pivot step may be summarized as follows: Replace the *pivot* $a(\neq 0)$ by $1/a$; multiply each remaining entry b in a's row by $1/a$ and each remaining entry c in a's column by $-1/a$; add to every other entry d the product $(-c/a)b$ of the new entry $-c/a$ in d's row and a's column and the old entry b in d's column and a's row. [It is important to verify that a pivot step is reversible: this assures solution-equivalence.]

The entire class of condensed tableaus for our illustrative program is exhibited in Table 2. There are eight tableaus corresponding to the eight cases in Table 1, as well as to the eight points of intersection in Figure 1. In the network (graph) of eight nodes (vertices) and eighteen branches (edges), each node corresponds to the immediately adjacent tableau and each branch joining two nodes to the existence of a pivot step (in either direction) from one to the other of the two corresponding tableaus (with appropriate rearrangement, if necessary, of rows and/or columns).

SIMPLEX METHOD

The network of tableaus for a general problem in $m + n$ variables x and $m + n$ variables y, with m "degrees of freedom" in the column system of equations and n "degrees of freedom" in the row system, may have as many as $(m + n)!/m!n!$ nodes (and $(m + n)!/2(m - 1)!(n - 1)!$ branches). For example, if $m = 50$ and $n = 150$, the number of nodes may be the order of 10^{47}. So, for a large-scale problem, it is clearly impracticable to construct the whole network of tableaus. Fortunately the ingenious Simplex Method (1947) of G. B. Dantzig provides an effective means of getting to an optimum by constructing only a very small part of the network.

Starting from a tableau, or node, which is y-feasible (or x-feasible), the Simplex Method proceeds through a succession of pivot steps, or branches, usually not more than $m + n$ in number, each of which preserves y-feasibility and makes v larger or at least not smaller (or preserves x-feasibility and makes u smaller or at least not larger) until a terminal tableau is attained that exhibits both x- and y-feasibility *or* shows such joint feasibility cannot exist.

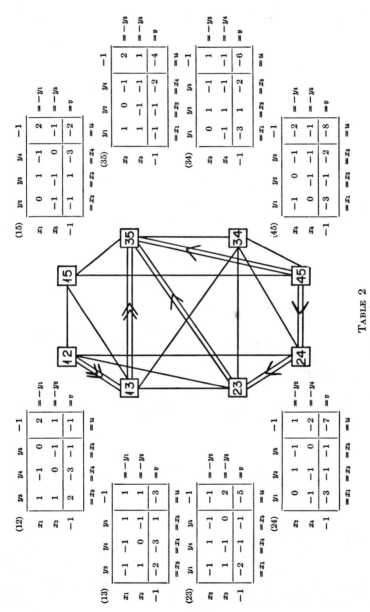

TABLE 2

Network of condensed tableaus (nodes) and pivot steps (branches)

For example, in Table 2, starting from the y-feasible schema 12 with $v = 1$, it is possible to pivot to the y-feasible tableau 13 with $v = 3$ and then to the y-feasible tableau 35 with $v = 4$, where x-feasibility is reached also and optimality thus secured. Alternatively, starting from the x-feasible tableau 45 with $u = 8$, it is possible to pivot to the x-feasible tableau 24 with $u = 7$, then to the x-feasible tableau 23 with $u = 5$, and finally to the x-feasible tableau 35 with $u = 4$, where y-feasibility is reached also and optimality thus secured; at the same time it is possible to pivot directly from 45 to 35. The double lines in Table 2 indicate the possibilities in our illustrative program for pivot steps preserving y-feasibility and increasing v *or* preserving x-feasibility and decreasing u. The single and double arrows indicate directions of decreasing u *or* increasing v, respectively; note that all these directions lead towards the node 35, at which optimality is secured. [It is possible to extend the Dantzig simplex algorithm to provide pivot steps from a node having neither x- nor y-feasibility, such as 15, to a y-feasible (or x-feasible) node, such as 12, or at least to a node having "less" infeasibility in y(or x).]

CONCLUSION

Until quite recently applications of mathematics occurred chiefly in the physical sciences and engineering, and drew mainly on calculus, differential equations and other parts of analysis. Now both the fields of application and the kinds of mathematics needed and used are vastly broader. Many advances in theory have come closer to their use in applications, both in time and in substance—especially in the newer areas.

Thus, linear programming which started in 1947 is already receiving wide usage in physical and social sciences, engineering, business and government operations. Today many people who do not consider themselves mathematicians need to understand the concepts and methods of linear programming in the regular course of their work. Conversely, the unsatisfied needs in applications, for both methods and methodology, have given direct stimulus to theoretical research.

This paper has presented a modest sample of a newly developing "combinatorial linear algebra." At the start our illustrative example seemed to be an *analytic* problem which might be treated by methods like standard ones of calculus. By focusing attention on the "key equation," however, and recognizing that a solution (if any exists) must arise within a finite class of simple possibilities, we were led to view our problem as a *combinatorial* one pertaining to a finite network of "condensed tableaus" as nodes and "pivot steps" as branches, within which the Dantzig simplex algorithm determines a remarkably short path to the desired optimum. Similar combinatorial structure can be identified in other linear "problems of organized complexity." It seems likely that, by the natural mathematical process of abstraction and generalization from various problems with common features, we will soon have a recognized body of "combinatorial (methods in) linear algebra".

Although the ideas are new, they are not difficult. Only tabular systems of linear equations and their pivot manipulation are needed for developing linear programming, as here portrayed, and for other salient topics in applied combinatorial mathematics. But while not difficult, the ideas can be far-reaching. For example, all the known theory concerning systems of linear inequalities can be developed in a thoroughly constructive fashion by an organized procedure based on tableaus and pivot steps.

Hence it now seems possible and indeed necessary to introduce some combinatorial mathematics, especially combinatorial aspects of linear algebra, at a fairly early stage in the mathematical curriculum and surely before major concentration at college. Linear programming has proved itself an effective vehicle for such instruction, adaptable even to secondary school mathematics. At the same time an understanding of linear programming is of increasing practical value in many occupations.

There is great challenge for the future in combinatorial mathematics.

REFERENCES

1. Weaver, Warren, "Science and Complexity," *American Scientist*, **36** (1948), 536.
2. Weyl, Hermann, *Philosophy of Mathematics and Natural Science*, Princeton, N.J.: Princeton University Press, 1949, p. 237; paperback, New York: Atheneum Press, 1963.
3. Dantzig, G. B., *Linear Programming and Extensions*, Princeton, N.J.: Princeton University Press, 1963.
4. Kemeny, J. G., J. L. Snell, and G. L. Thompson, *Introduction to Finite Mathematics*, 3rd edition, Chap. 7, Englewood Cliffs, N.J.: Prentice-Hall, 1974.
5. Gewirtz, A., H. Sitomer, and A. W. Tucker, *Constructive Linear Algebra*, Chap. 1, Englewood Cliffs, New Jersey: Prentice-Hall, 1974.
6. Singleton, R. R., and W. F. Tyndall, *Games and Programs: Mathematics for Modeling*, Chaps. 7–11, San Francisco: Freeman, 1974.

COMPLEMENTARY PIVOT THEORY OF MATHEMATICAL PROGRAMMING

Richard W. Cottle[1]
George B. Dantzig[2]

1. FORMULATION

Linear programming, quadratic programming, and bimatrix (two-person, nonzero-sum) games lead to the consideration of the following *Fundamental Problem:*[3] Given a real p-vector q and a

1. Research partially supported by National Science Foundation Grant GP-3739.

2. Research partially supported by U.S. Army Research Office Contract No. DAHC04-67-C-0028, Office of Naval Research, Contract ONR-N-00014-67-A-0112-0011, U.S. Atomic Energy Commission, Contract No. AT(04-3)-326 PA # 18, and National Science Foundation Grant GP 6431.

3. The fundamental problem can be extended *from* p sets each consisting of a pair of variables only one of which can be nonbasic *to* k sets of several variables each, only one of which can be nonbasic. To be specific, consider a system $w = q + Nz$, $w \geqq 0$, $z \geqq 0$ where N is a $p \times k$ matrix ($k \leqq p$) and the variables $w_1, \cdots, w_p$ are partitioned into k nonempty sets $S_l, l = 1, \cdots, k$. Let $T_l = S_l \cup \{z_l\}$, $l = 1, \cdots, k$. We seek a solution of the system in which exactly one member of each set T_l is nonbasic. (The fundamental problem is of this form where $k = p$ and $T_l = \{w_l, z_l\}$.) The underlying idea of Lemke's approach §2 applies here. For example, it can be shown that this problem has a solution when $N > 0$.

real $p \times p$ matrix M, find vectors w and z which satisfy the conditions[4]

(1) $$w = q + Mz, \quad w \geqq 0, z \geqq 0$$

(2) $$zw = 0.$$

The remainder of this section is devoted to an explanation of why this is so. (There are other fields in which this fundamental problem arises—see for example [6] and [13]—but we do not treat them here.) §§2 and 3 are concerned with constructive procedures for solving the fundamental problem under various assumptions on the data q and M.

Consider first linear programs in the symmetric primal-dual form due to J. von Neumann [20].

Primal linear program: Find a vector x and minimum $\bar{z}$ such that

(3) $$Ax \geqq b, \quad x \geqq 0, \quad \bar{z} = cx.$$

Dual linear program: Find a vector y and maximum $\underset{\sim}{z}$ such that

(4) $$yA \leqq c, \quad y \geqq 0, \quad \underset{\sim}{z} = yb.$$

The duality theorem of linear programming [3] states that min $\bar{z}$ = max $\underset{\sim}{z}$ when the primal and dual systems (3) and (4), respectively, are consistent or—in mathematical programming parlance—"feasible." Since

$$\underset{\sim}{z} = yb \leqq yAx \leqq cx = \bar{z}$$

for all primal-feasible x and dual-feasible y, one seeks such solutions for which

(5) $$yb = cx.$$

---

4. In general, capital *italic* letters denote matrices while vectors are denoted by lower case *italic* letters. Whether a vector is a row or a column will always be clear from the context, and consequently we dispense with transpose signs on vectors. In (2), for example, zw represents the scalar product of z (row) and w (column). The superscript ^{T} indicates the transpose of the matrix to which it is affixed.

The inequality constraints of the primal and dual problems can be converted to equivalent systems of equations in nonnegative variables through the introduction of nonnegative "slack" variables. Jointly, the systems (3) and (4) are equivalent to

$$(6) \qquad Ax - v = b, \qquad v \geqq 0, x \geqq 0,$$

$$A^T y + u = c, \qquad u \geqq 0, y \geqq 0,$$

and the linear programming problem becomes one of finding vectors u, v, x, y such that

$$(7) \qquad \begin{pmatrix} u \\ v \end{pmatrix} = \begin{pmatrix} c \\ -b \end{pmatrix} + \begin{pmatrix} 0 & -A^T \\ A & 0 \end{pmatrix} \begin{pmatrix} x \\ y \end{pmatrix} \qquad \begin{array}{l} u \geqq 0, v \geqq 0, \\ x \geqq 0, y \geqq 0, \end{array}$$

and (by (5))

$$(8) \qquad xu + yv = 0.$$

The definitions

$$(9) \quad w = \begin{pmatrix} u \\ v \end{pmatrix}, \quad q = \begin{pmatrix} c \\ -b \end{pmatrix}, \quad M = \begin{pmatrix} 0 & -A^T \\ A & 0 \end{pmatrix}, \quad z = \begin{pmatrix} x \\ y \end{pmatrix}$$

establish the correspondence between (1), (2) and (3), (4).

The *quadratic programming problem* is typically stated in the following manner: Find a vector x and minimum $\bar{z}$ such that

$$(10) \qquad Ax \geqq b, \qquad x \geqq 0, \qquad \bar{z} = cx + \tfrac{1}{2}xDx.$$

In this formulation, the matrix D may be assumed to be symmetric. The minimand $\bar{z}$ is a globally convex function of x if and only if the quadratic form xDx (or matrix D) is positive semidefinite, and when this is the case, (10) is called the *convex quadratic programming problem*. It is immediate that when D is the zero matrix, (10) reduces to the linear program (3). In this sense, the linear programming problem is a special case of the quadratic programming problem.

For any quadratic programming problem (10), define u and v by

$$(11) \qquad u = Dx - A^T y + c, \qquad v = Ax - b.$$

A vector x° yields minimum $\bar{z}$ only if there exists a vector y° and vectors u°, v° given by (11) for $x = x^\circ$ satisfying

$$(12) \qquad x^\circ \geqq 0, \qquad u^\circ \geqq 0, \qquad y^\circ \geqq 0, \qquad v^\circ \geqq 0,$$
$$x^\circ u^\circ = 0, \qquad y^\circ v^\circ = 0.$$

These *necessary conditions* for a minimum in (10) are a direct consequence of a theorem of H. W. Kuhn and A. W. Tucker [14]. It is well known—and not difficult to prove from first principles— that (12), known as the Kuhn-Tucker conditions, are also *sufficient* in the case of convex quadratic programming. By direct substitution, we have for any feasible vector x,

$$\begin{aligned}
\bar{z} - \bar{z}^0 &= c(x - x^0) + \tfrac{1}{2}xDx - \tfrac{1}{2}x^0Dx^0 \\
&= u^\circ(x - x^\circ) + y^\circ(v - v^\circ) + \tfrac{1}{2}(x - x^\circ)D(x - x^\circ) \\
&= u^\circ x + y^\circ v + \tfrac{1}{2}(x - x^\circ)D(x - x^\circ) \geqq 0
\end{aligned}$$

which proves the sufficiency of conditions (12) for a minimum in the convex case.

Thus, the problem of solving a quadratic program leads to a search for solution of the system

$$(13) \qquad u = Dx - A^T y + c \qquad x \geqq 0, y \geqq 0,$$
$$v = Ax - b \qquad u \geqq 0, v \geqq 0,$$

$$(14) \qquad\qquad xu + yv = 0.$$

The definitions

$$(15) \quad w = \begin{pmatrix} u \\ v \end{pmatrix}, \qquad q = \begin{pmatrix} c \\ -b \end{pmatrix}, \qquad M = \begin{pmatrix} D & -A^T \\ A & 0 \end{pmatrix}, \qquad z = \begin{pmatrix} x \\ y \end{pmatrix}$$

establish (13), (14) as a problem of the form (1), (2).

Dual of a convex quadratic program. From (15) one is led naturally to the consideration of a matrix

$$M = \begin{pmatrix} D & -A^T \\ A & E \end{pmatrix}$$

wherein E, like D, is positive semidefinite. It is shown in [1] that the

Primal quadratic program: Find x, y and minimum $\bar{z}$ such that

$$(16) \quad Ax + Ey \geqq b, \quad x \geqq 0, \quad \bar{z} = cx + \tfrac{1}{2}(xDx + yEy)$$

has the associated

Dual quadratic program: Find x, y and maximum $\underline{z}$ such that

$$(17) \quad -Dx + A^T y \leqq c, \quad y \geqq 0, \quad \underline{z} = by - \tfrac{1}{2}(xDx - yEy)$$

All the results of duality in linear programming extend to these problems, and indeed they are jointly solvable if either is solvable. When $E = 0$, the primal problem is just (10) for which W. S. Dorn [5] first established the duality theory later extended in [1]. When both D and E are zero matrices, this dual pair (16), (17) reduces to the dual pair of linear programs (3), (4).

REMARKS. (a) The minimand in (10) is strictly convex if and only if the quadratic form xDx is positive definite. Any *feasible* strictly convex quadratic program has a unique minimizing solution x°.

(b) When D and E are positive semidefinite (the case of convex quadratic programming), so is

$$M = \begin{pmatrix} D & -A^T \\ A & E \end{pmatrix}.$$

A *bimatrix* (or two-person nonzero-sum) *game*, $\Gamma(A, B)$, is given by a pair of $m \times n$ matrices A and B. One party, called the *row player*, has m pure strategies which are identified with the rows of A. The other party, called the *column player*, has n pure strategies which correspond to the columns of B. If the row player uses his ith pure strategy and the column player uses his jth pure strategy, then their respective *losses* are defined as a_{ij} and b_{ij}, respectively. Using *mixed strategies*

$$x = (x_1, \cdots, x_m) \geqq 0, \qquad \sum_{i=1}^{m} x_i = 1,$$

$$y = (y_1, \cdots, y_n) \geqq 0, \qquad \sum_{j=1}^{n} y_j = 1,$$

their expected losses are xAy and xBy, respectively. (A component in a mixed strategy is interpreted as the probability with which the player uses the corresponding pure strategy.)

A pair (x°, y°) of mixed strategies is a *Nash* [19] *equilibrium point* of $\Gamma(A, B)$ if

$$x^\circ A y^\circ \leqq x A y^\circ \quad \text{all mixed strategies } x,$$

$$x^\circ B y^\circ \leqq x^\circ B y \quad \text{all mixed strategies } y.$$

It is evident (see for example [15]) that if (x°, y°) is an equilibrium point of $\Gamma(A, B)$, then it is also an equilibrium point for the game $\Gamma(A', B')$ in which

$$A' = [a_{ij} + K], \qquad B' = [b_{ij} + L]$$

where K and L are arbitrary scalars. Hence there is no loss of generality in assuming that $A > 0$ and $B > 0$, and we shall make this assumption hereafter.

Next, by letting e_k denote the k-vector all of whose components are unity, it is easily shown that (x°, y°) is an equilibrium point of $\Gamma(A, B)$ if and only if

(18) $(x^\circ A y^\circ) e_m \leqq A y^\circ \qquad (A > 0),$

(19) $(x^\circ B y^\circ) e_n \leqq B^T x^\circ \qquad (B > 0).$

This characterization of an equilibrium point leads to a theorem which relates the equilibrium-point problem to a system of the form (1), (2). For $A > 0$ and $B > 0$, if u^*, v^*, x^*, y^* is a solution of the system

(20) $u = Ay - e_m, \qquad u \geqq 0, y \geqq 0,$

$$v = B^T x - e_n, \qquad v \geqq 0, x \geqq 0,$$

(21) $xu + yv = 0$

then

$$(x^\circ, y^\circ) = (x^*/x^*e_m, y^*/y^*e_n)$$

is an equilibrium point of $\Gamma(A, B)$. Conversely, if (x°, y°) is an equilibrium point of $\Gamma(A, B)$ then

$$(x^*, y^*) = (x^\circ/x^\circ By^\circ, y^\circ/x^\circ Ay^\circ)$$

is a solution of (20), (21). The latter system is clearly of the form (1), (2), where

$$w = \begin{pmatrix} u \\ v \end{pmatrix}, \qquad q = \begin{pmatrix} -e_m \\ -e_n \end{pmatrix}, \qquad M = \begin{pmatrix} 0 & A \\ B^T & 0 \end{pmatrix}, \qquad z = \begin{pmatrix} x \\ y \end{pmatrix}.$$

Notice that the assumption $A > 0, B > 0$ precludes the possibility of the matrix M above belonging to the positive semidefinite class.

The existence of an equilibrium point for $\Gamma(A, B)$ was established by J. Nash [19] whose proof employs the Brouwer Fixed-Point Theorem. Recently, an elementary constructive proof was discovered by C. E. Lemke and J. T. Howson, Jr. [15].

2. LEMKE'S ITERATIVE SOLUTION OF THE FUNDAMENTAL PROBLEM

This section is concerned with the iterative technique of Lemke and Howson for finding equilibrium points of bimatrix games which was later extended by Lemke to the fundamental problem (1), (2). We introduce first some terminology common to the subject of this section and the next. Consider the system of linear equations

$$(22) \qquad w = q + Mz$$

where, for the moment, the p-vector q and the $p \times p$ matrix M are arbitrary. Both w and z are p-vectors.

For $i = 1, \cdots, p$ the corresponding variables z_i and w_i are called *complementary* and each is the *complement* of the other. A *complementary solution* of (22) is a pair of vectors satisfying (22) and

$$(23) \qquad z_i w_i = 0, \qquad i = 1, \cdots, p.$$

Notice that a solution $(w; z)$ of (1), (2) is a nonnegative complementary solution of (22). Finally, a solution of (22) will be called *almost-complementary* if it satisfies (23) except for one value of i, say $i = \beta$. That is, $z_\beta \neq 0$, $w_\beta \neq 0$.

In general, the procedure assumes as *given* an extreme point of the convex set

$$Z = \{z \mid w = q + Mz \geqq 0, z \geqq 0\}$$

which also happens to be the endpoint of an almost-complementary ray (unbounded edge) of Z. Each point of this ray satisfies (23) but for one value of i, say β. It is not always easy to find such a starting point for an arbitrary M. Yet there are two important realizations of the fundamental problem which can be so initiated. The first is the bimatrix game case to be discussed soon; the second is the case where an entire column of M is positive. The latter property can always be *artificially* induced by augmenting M with an additional positive column; as we shall see, this turns out to be a useful device for initiating the procedure with a general M.

Each iteration corresponds to motion from an extreme point P_i along an edge of Z all points of which are almost-complementary solutions of (22). If this edge is bounded, an adjacent extreme point P_{i+1} is reached which is either complementary or almost-complementary. The process terminates if (i) the edge is unbounded (a ray), (ii) P_{i+1} is a previously generated extreme point, or (iii) P_{i+1} is a complementary extreme point.

Under the assumption of nondegeneracy, the extreme points of Z are in one-to-one correspondence with the *basic feasible solutions* of (22). (See [3].) Still under this assumption, a *complementary basic feasible solution* is one in which the complement of each basic variable is nonbasic. The goal is to obtain a basic feasible solution with such a property. In an almost-complementary basic feasible of (23), there will be exactly one index, say β, such that both w_β and z_β are basic variables. Likewise, there will be exactly one index, say ν, such that both w_ν and z_ν are nonbasic variables.[5]

---

5. C. van de Panne and A. Whinston [21] have used the appropriate terms *basic* and *nonbasic pair* for $\{w_\beta, z_\beta\}$ and $\{w_\nu, z_\nu\}$ respectively.

An almost-complementary edge is generated by holding all nonbasic variables at value zero and increasing either z_ν or w_ν of the nonbasic pair z_ν, w_ν. There are consequently *exactly two* almost-complementary edges associated with an almost-complementary extreme point (corresponding to an almost-complementary basic feasible solution).

Suppose that z_ν is the nonbasic variable to be increased. The values of the basic variables will change linearly with the changes in z_ν. For sufficiently small positive values of z_ν, the almost-complementary solution remains feasible. This is a consequence of the nondegeneracy assumption. But in order to retain feasibility, the values of the basic variables must be prevented from becoming negative.

If the value of z_ν can be made arbitrarily large without forcing any basic variable to become negative, then a *ray* is generated. In this event, the process terminates. However, if some basic variable *blocks* the increase of z_ν (i.e. vanishes for a positive value of z_ν), then a new basic solution is obtained which is either complementary or almost-complementary. A complementary solution occurs only if a member of the basic pair blocks z_ν. A new almost-complementary extreme point solution is obtained if the blocking occurs otherwise. In the complementary case, we have the desired result: a complementary basic feasible solution. In the almost-complementary case, the nondegeneracy assumption guarantees the uniqueness of the blocking variable. It will become nonbasic in place of z_ν and its index becomes the new value of ν.

The complementary rule. The complement of the (now nonbasic) blocking variable—or equivalently put, the other member of the "new" nonbasic pair—is the next nonbasic variable to be increased. The procedure consists of the iteration of these steps. The generated sequence of almost-complementary extreme points and edges is called an *almost-complementary path*.

THEOREM 1: *Along an almost-complementary path, the only almost-complementary basic feasible solution which can re-occur is the initial one.*

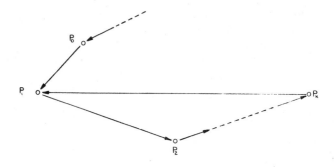

Proof: We assume that all basic feasible solutions of (22) are nondegenerate. (This can be assured by any of the standard lexicographic techniques [3] for resolving the ambiguities of degeneracy.) Suppose, contrary to the assertion of the theorem, that the procedure generates a sequence of almost-complementary basic feasible solutions in which a term other than the first one (P_0 in the figure below) is repeated (say P_1). By the nondegeneracy assumption, the extreme points of Z are in one-one correspondence with basic feasible solutions of (22). Let P_2 denote the successor of P_1 and let P_k denote the second predecessor to P_1 namely the one along the path just before the return to P_1. The extreme points P_0, P_2, P_k are distinct and each is adjacent to P_1 along an almost-complementary edge. But there are only *two* such edges at P_1. This contradiction completes the proof.

We can immediately state the

COROLLARY: *If the almost-complementary path is initiated at the endpoint of an almost-complementary ray, the procedure must terminate either in a different ray or a complementary basic feasible solution.*

It is easy to show by examples that starting from an almost-complementary basic feasible solution which is *not* the endpoint of an almost-complementary ray, the procedure *can* return to the initial point regardless of the existence or nonexistence of a solution to (1), (2).

EXAMPLE 1. The set Z associated with

$$q = \begin{pmatrix} 1 \\ -1 \\ 3 \end{pmatrix}, \qquad M = \begin{pmatrix} 0 & 0 & 0 \\ 1 & 0 & 0 \\ -1 & -1 & -1 \end{pmatrix}$$

is nonempty and bounded. It is clear that no solution of (1) can also satisfy (2) since $z_1 w_1 > 0$. Let the extreme point corresponding to the solution $w = (1, 0, 0)$, $z = (1, 0, 2)$ be the initial point of a path which begins by increasing z_2. This will return to the initial extreme point after 4 iterations.

EXAMPLE 2. The set Z associated with

$$q = \begin{pmatrix} 1 \\ -1 \\ 3 \\ 1 \end{pmatrix}, \qquad M = \begin{pmatrix} 0 & 0 & 0 & 0 \\ 1 & 0 & 0 & 1 \\ -1 & -1 & -1 & -1 \\ 0 & 0 & 0 & -1 \end{pmatrix}$$

is likewise nonempty and bounded. The corresponding fundamental problem (1), (2) has a complementary solution $w = (1, 0, 1, 0)$, $z = (0, 1, 0, 1)$. Yet by starting at $w = (1, 2, 0, 1)$, $z = (3, 0, 0, 0)$ and increasing z_3, the method generates a path which returns to its starting point after 4 iterations.

Furthermore, even if the procedure is initiated from an extreme point at the end of an almost-complementary ray, termination in a ray is possible whether or not the fundamental problem has a solution.

EXAMPLE 3. Given the data

$$q = \begin{pmatrix} 1 \\ -1 \\ 3 \\ 1 \end{pmatrix}, \qquad M = \begin{pmatrix} 0 & 0 & 0 & 1 \\ 1 & 0 & 0 & 1 \\ 1 & -1 & 1 & 1 \\ 0 & 0 & 0 & -1 \end{pmatrix}$$

the point of Z corresponding to $w = (1, 0, 4, 1)$, $z = (1, 0, 0, 0)$ is at the end of an almost-complementary ray, $w = (1, w_2, 4 + w_2, 1)$, $z = (1 + w_2, 0, 0, 0)$. Moving along the edge generated by increasing z_2 leads to a new almost-complementary extreme point at which the required increase of z_3 is unblocked, so that the process terminates in a ray, and yet the fundamental problem is solved by $w = (2, 0, 3, 0)$, $z = (0, 1, 0, 1)$.

EXAMPLE 4. In the problem with

$$q = \begin{pmatrix} 1 \\ -1 \end{pmatrix}, \qquad M = \begin{pmatrix} 0 & 0 \\ 1 & -1 \end{pmatrix}$$

the inequalities (1) have solutions, but none of them satisfy (2). The point corresponding to $(w; z) = (1, 0; 1, 0)$ is at the end of an almost-complementary ray $w = (1, w_2)$, $z = (1 + w_2, 0)$. When z_2 is increased, it is not blocked, and the process terminates in a ray.

Consequences of termination in a ray. In this geometrical approach to the fundamental problem, it is useful to interpret algebraically the meaning of termination in an almost-complementary ray. This can be achieved by use of a standard result in linear inequality theory [11], [3].

LEMMA: *If* $(w^*; z^*)$ *is an almost-complementary basic feasible solution of* (22), *and* $(w^*; z^*)$ *is incident to an almost-complementary ray, there exist p-vectors* w^h, z^h *such that*

$$(24) \qquad w^h = Mz^h, \qquad w^h \geqq 0, \qquad z^h \geqq 0, \qquad z^h \neq 0$$

and points along the almost-complementary ray are of the form

$$(25) \qquad\qquad (w^* + \lambda w^h, z^* + \lambda z^h), \qquad \lambda \geqq 0$$

and satisfy

$$(26) \qquad (w_i^* + \lambda w_i^h)(z_i^* + \lambda z_i^h) = 0 \quad \textit{for all } \lambda \geqq 0, \quad \textit{and all } i \neq \beta.$$

THEOREM 2: *If* $M > 0$, (22) *has a complementary basic feasible solution for any vector* q.

Proof: Select $w_1, \cdots, w_p$ as the basic variables in (22). We may assume that $q \ngeqq 0$ for otherwise $(w; z) = (q; 0)$ immediately solves the problem. A starting ray of feasible almost-complementary solutions is generated by taking a sufficiently large value of any nonbasic variable, say z_1. Reduce z_1 toward zero until it reaches a value $z_1^\circ \geqq 0$ at which a unique basic variable (assuming nondegeneracy) becomes zero. An extreme point has then been reached.

The procedure has been initiated in the manner described by the corollary above, and consequently the procedure must terminate either in a complementary basic feasible solution or in an almost-complementary ray after some basic feasible solution $(w^*; z^*)$ is reached. We now show that the latter cannot happen. For if it does, conditions (24)–(26) of the lemma obtain with $\beta = 1$. Since $M > 0$ and $z^h \geqq 0$, this implies $w^h > 0$. Hence by (26), $z_i^* = z_i^h = 0$ for all $i \neq 1$. Hence the only variables which change with λ are z_1 and the components of w. Therefore the final generated ray is the same as the initiating ray, which contradicts the corollary.

THEOREM 3: *A bimatrix game $\Gamma(A, B)$ has an extreme equilibrium point.*

Proof: Initiate the algorithm by choosing the smallest positive value of x_1, say x_1°, such that

$$(27) \qquad v = -e_n + B_1^T x_1^\circ \geqq 0$$

where B_1^T is the first column of B^T. With

$$v^\circ = -e_n + B_1^T x_1^\circ$$

it follows (assuming nondegeneracy) that v° has exactly one zero component, say the rth. The ray is generated by choosing as basic variables x_1 and all the slack variables u, v except for v_r. The complement of v_r, namely y_r, is chosen as the nonbasic variable to increase indefinitely. For sufficiently large values of y_r, the basic variables are all nonnegative and the ray so generated is complementary except possibly $x_1 u_1$ might not equal 0. Letting y_r decrease toward zero, the initial extreme point is obtained for some positive value of y_r.

If the procedure does not terminate in an equilibrium point, then by the corollary, it terminates in an almost-complementary ray. The latter implies the existence of a class of almost-complementary solutions of the form[6]

$$(28) \qquad \begin{pmatrix} u^* + \lambda u^h \\ v^* + \lambda v^h \end{pmatrix} = \begin{pmatrix} -e_m \\ -e_n \end{pmatrix} + \begin{pmatrix} 0 & A \\ B^T & 0 \end{pmatrix} \begin{pmatrix} x^* + \lambda x^h \\ y^* + \lambda y^h \end{pmatrix}.$$

$$(29) \qquad (u_i^* + \lambda u_i^h)(x_i^* + \lambda x_i^h) = 0 \quad \text{all } i \neq 1$$

$$(30) \qquad (v_j^* + \lambda v_j^h)(y_j^* + \lambda y_j^h) = 0 \quad \text{all } j$$

$$\left. \right\} \text{ all } \lambda \geqq 0.$$

Assume first that $x^h \neq 0$. Then $v^h = B^T x^h > 0$. By (30), $y_j^* + \lambda y_j^h = 0$ for all j and all $\lambda \geqq 0$. But then $u^* + \lambda u^h = -e_m < 0$, a contradiction. Assume next that $y^h \neq 0$ and $x^h = 0$. Then $u^h = Ay^h > 0$. By (29), $x_i^* = 0$ for all $i \neq 1$; and $x_i^h = 0$ for all i. Hence $v^h = B^T x^h = 0$ and v^* is the same as v defined by (27) since x_1 must be at the smallest value in order that (u^*, v^*, x^*, y^*) be an extreme-point solution. By the nondegeneracy assumption, only $v_r^* = 0$, and $v_j^* > 0$ for all $j \neq r$. Hence (30) implies $y_j^* + \lambda y_j^h = 0$ for all $j \neq r$. It is now clear that the postulated terminating ray is the original ray. This furnishes the desired contradiction. The algorithm must terminate in an equilibrium point of the bimatrix game $\Gamma(A, B)$.

A modification of almost-complementary basic sets. Consider the system of equations

$$(31) \qquad\qquad w = q + e_p z_0 + Mz$$

where z_0 represents an "artificial variable" and e_p is a p-vector $(1, \cdots, 1)$. It is clear that (31) always has nonnegative solutions. A solution of (31) is called *almost complementary* if $z_i w_i = 0$ for $i = 1, \cdots, p$ and is *complementary* if, in addition, $z_0 = 0$. (See [16, p. 685] where a different but equivalent definition is given.) In this case, let

$$Z_0 = \{(z_0, z) \mid w = q + e_p z_0 + Mz \geqq 0, z_0 \geqq 0, z \geqq 0\}.$$

We consider the almost-complementary ray generated by suf-

---

6. The notational analogy with the previously studied case $M > 0$ is obvious.

ficiently large z_0. The variables $w_1, \cdots, w_p$ are initially basic while $z_0, z_1, \cdots, z_p$ are nonbasic variables. For a sufficiently large value of z_0, say z_0^+,

$$w^+ = q + e_p z_0^+ > 0.$$

As z_0 decreases toward zero, the basic variables w_i decrease. An initial extreme point is reached when z_0 attains the minimum value z_0° for which $w = q + e_p z_0 \geq 0$. If $z_0^\circ = 0$, then $q \geq 0$; this is the trivial case for which no algorithm is required. If $z_0^\circ > 0$, some unique basic variable, say w_r has reached its lower bound 0. Then z_0 becomes a basic variable in place of w_r and we have $\nu = r$. Next, z_r, the complement of w_r, is to be increased.

The remaining steps of the procedure are now identical to those in the preceding algorithm. After a blocking variable becomes basic, its complement is increased until either a basic variable blocks the increase (by attaining its lower bound 0) or else an almost-complementary ray is generated. There are precisely two forms of termination. One is in a ray as just described; the other is in the reduction of z_0 to the value 0 and hence the attainment of a complementary basic feasible solution of (31), i.e. a solution of (1), (2).

Interest now centers on the meaning of termination in an almost-complementary ray solution of (31). *For certain classes of matrices, the process described above terminates in an almost-complementary ray if and only if the original system (1) has no solution.* In the remainder of this section, we shall amplify the preceding statement.

If termination in an almost-complementary ray occurs after the process reaches a basic feasible solution $(w^*; z_0^*, z^*)$ corresponding to an extreme point of Z_0, then there exists a nonzero vector $(w^h; z_0^h; z^h)$ such that

$$(32) \qquad w^h = e_p z_0^h + M z^h, \qquad (w^h; z_0^h; z^h) \geq 0.$$

Moreover for every $\lambda \geq 0$,

$$(33) \qquad (w^* + \lambda w^h) = q + e_p(z_0^* + \lambda z_0^h) + M(z^* + \lambda z^h)$$

and

$$(34) \qquad (w_i^* + \lambda w_i^h)(z_i^* + \lambda z_i^h) = 0, \qquad i = 1, \cdots, p.$$

The case $z^h = 0$ is ruled out, for otherwise $z_0^h > 0$ and then $w^h > 0$ because $(w^h; z_0^h, z^h) \neq 0$. Now if $w^h > 0$, (34) implies $z^* + \lambda z^h = z^* = 0$. This, in turn, implies that the ray is the original one which is not possible.

Furthermore, it follows from the almost-complementarity of solutions along the ray that

$$(35) \quad z_i^* w_i^* = z_i^* w_i^h = z_i^h w_i^* = z_i^h w_i^h = 0, \qquad i = 1, \cdots, p.$$

The individual equations of the system (32) are of the form

$$(36) \qquad w_i^h = z_0^h + (Mz^h)_i, \qquad i = 1, \cdots, p.$$

Multiplication of (36) by z_i^h leads, via (35), to

$$(37) \qquad 0 = z_i^h z_0^h + z_i^h (Mz^h)_i, \qquad i = 1, \cdots, p,$$

from which we conclude

THEOREM 4: *Termination in a ray implies there exists a nonzero nonnegative vector z^h such that*

$$(38) \qquad z_i^h (Mz^h)_i \leqq 0, \qquad i = 1, \cdots, p.$$

At this juncture, two large classes of matrices M will be considered. For the first class, we show that termination in a ray implies the *inconsistency* of the system (1). For the second class, we will show that termination in a ray cannot occur, so that for this class of matrices (1), (2) always has a solution regardless of what q is.

The first class mentioned above was introduced by Lemke [16]. These matrices, which we shall refer to as *copositive plus*, are required to satisfy the two conditions,

$$(39) \qquad uMu \geqq 0 \qquad \text{for all} \quad u \geqq 0,$$

$$(40) \quad (M + M^T)u = 0 \qquad \text{if} \quad uMu = 0 \quad \text{and} \quad u \geqq 0.$$

Matrices satisfying conditions (39) alone are known in the literature as *copositive* (see [18], [12]). To our knowledge, there is no reference other than [16] on copositive matrices satisfying the condition (40). However, the class of such matrices is large and includes

(i) all *strictly copositive matrices*, i.e. those for which $uMu > 0$ when $0 \neq u \geq 0$,

(ii) all positive semidefinite matrices, i.e. those for which $uMu \geq 0$ for all u.

Positive matrices are obviously strictly copositive while positive definite matrices are both positive semidefinite and strictly copositive. Furthermore, it is possible to "build" matrices satisfying (39) and (40) out of smaller ones. For example, if M_1 and M_2 are matrices satisfying (39) and (40) then so is the block-diagonal matrix

$$M = \begin{pmatrix} M_1 & 0 \\ 0 & M_2 \end{pmatrix}.$$

Moreover, if M satisfies (39) and (40) and S is any skew-symmetric matrix (of its order), then $M + S$ satisfies (39) and (40). Consequently, block matrices such as

$$M = \begin{pmatrix} M_1 & -A^T \\ A & M_2 \end{pmatrix}$$

satisfy (39) and (40) if and only if M_1 and M_2 do too. However, as Lemke [16], [17] has pointed out, the matrices encountered in the bimatrix game problem with $A > 0$ and $B > 0$ need not satisfy (40). The Lemke-Howson iterative procedure for bimatrix games was given earlier in this section. If applied to bimatrix games, the modification just given always terminates in a ray after just one iteration, as can be verified by taking any example.

The second class, consisting of matrices having *positive principal minors*, has been studied by numerous investigators; see for example, [2], [4], [8], [9], [10], [22], [24]. In the case of symmetric matrices, those with positive principal minors are positive definite. But the equivalence breaks down in the nonsymmetric situation. Nonsymmetric matrices with positive principal minors need not be positive definite. For example, the matrix

$$\begin{pmatrix} 2 & -7 \\ -1 & 4 \end{pmatrix}$$

has positive principal minors but is indefinite and not copositive. However, positive definite matrices are a subset of those with positive principal minors. (See, e.g. [2].)

We shall make use of the fact that $w = q + Mz$, $(w; z) \geqq 0$ has no solution if there exists a vector v such that

(41) $\qquad vM \leqq 0, \qquad vq < 0, \qquad v \geqq 0$

for otherwise, $0 \leqq vw = vq + vMz < 0$, a contradiction. Indeed, it is a consequence of J. Farkas' theorem [7] that (1) has no solution if and only if there exists a solution of (41).

THEOREM 5: *Let M be copositive plus. If the iterative procedure terminates in a ray, then* (1) *has no solution.*

Proof: Termination in a ray means that a basic feasible solution $(w^*; z_0^*, z^*)$ will be reached at which conditions (32)–(34) hold and also

(42) $\qquad 0 = z^h w^h = z^h e_p z_0^h + z^h M z^h.$

Since M is copositive and $z^h \geqq 0$, both terms on the right side of (42) are nonnegative, hence both are zero. The scalar $z_0^h = 0$ because $z^h e_p > 0$. The vanishing of the quadratic form $z^h M z^h$ means

$$Mz^h + M^T z^h = 0.$$

But by (32), $z_0^h = 0$ implies that $w^h = Mz^h \geqq 0$, whence, $M^T z^h \leqq 0$ or, what is the same thing, $z^h M \leqq 0$. Next, by (35),

$$0 = z^* w^h = z^* M z^h = z^*(-M^T z^h) = -z^h M z^*$$

and we obtain again by (35)

$$0 = z^h w^* = z^h q + z^h e_p z_0^* + z^h M z^* = z^h q + z^h e_p z_0^*.$$

It follows that $z^h q < 0$ because $z^h e_p z_0^* > 0$. The conditions (1) are therefore inconsistent because $v = z^h$ satisfies (41).

COROLLARY: *If M is strictly copositive, the process terminates in a complementary basic feasible solution of* (31).

Proof: If not, the proof of Theorem 5 would imply the existence of a vector z^h satisfying $z^h M z^h = 0$, $0 \neq z^h \geq 0$ which contradicts the strict copositivity of M.

This corollary clearly generalizes Theorem 1. We now turn to the matrices M having positive principal minors.

THEOREM 6: *If M has positive principal minors, the process terminates in a complementary basic solution of* (31) *for any q.*

Proof: We have seen that termination in a ray implies the existence of a nonzero vector z^h satisfying the inequalities (38). However, Gale and Nikaido [10, Theorem 2] have shown that matrices with positive principal minors are characterized by the impossibility of this event. Hence termination in a ray is not a possible outcome for problems in which M has positive principal minors.

We can even improve upon this.

THEOREM 7: *If M has the property that for each of its principal submatrices $\tilde{M}$, the system*

$$\tilde{M}\tilde{z} \leq 0, \qquad 0 \neq \tilde{z} \geq 0$$

has no solution, then the process terminates in a complementary basic solution of (31) *for any q.*

Proof: Suppose the process terminates in a ray. From the solution $(w^h; z_0^h, z^h)$ of the homogeneous system (32), define the vector $\tilde{w}^h$ of components of w^h for which the corresponding component of $z^* + z^h$ is positive. Then by (34) $\tilde{w}^h = 0$. Let $\tilde{z}^h$ be the vector of corresponding components in z^h. Clearly $0 \neq \tilde{z}^h \geq 0$, since $0 \neq z^h \geq 0$ and any positive component of z^h is a positive component of $\tilde{z}^h$ by definition of $\tilde{w}^h$. Let $\tilde{M}$ be the corresponding principal submatrix of M. Since $\tilde{M}$ is a matrix of order $k \geq 1$ we may write

$$0 = \tilde{w}^h = e_k z_0^h + \tilde{M}\tilde{z}^h.$$

Hence

$$\tilde{M}\tilde{z}^h \leq 0, \qquad 0 \neq \tilde{z}^h \geq 0,$$

which is a contradiction.

3. THE PRINCIPAL PIVOTING METHOD

We shall now describe an algorithm proposed by the authors [4] which predates that of Lemke. It evolved from a quadratic programming algorithm of P. Wolfe [26] who was the first to use a type of complementary rule for pivot choice. Our method is applicable to matrices M that have positive principal minors (in particular to positive definite matrices) and after a minor modification, to positive semidefinite matrices.

In Lemke's procedure for general M, an artificial variable z_0 is introduced in order to obtain feasible almost-complementary solutions for the augmented problem. In our approach, only variables of the original problem are used, but these can take on initially negative as well as nonnegative values.

A *major cycle* of the algorithm is initiated with the complementary basic solution $(w; z) = (q; 0)$. If $q \geq 0$, the procedure is immediately terminated. If $q \ngeq 0$, we may assume (relabeling if necessary) that $w_1 = q_1 < 0$. An almost-complementary path is generated by increasing z_1, the complement of the selected negative basic variable. For points along the path, $z_i w_i = 0$ for $i \neq 1$.

Step I. Increase z_1 until it is blocked by a positive basic variable decreasing to zero or by the negative w_1 increasing to zero.

Step II. Make the blocking variable nonbasic by pivoting its complement into the basic set. The major cycle is terminated if w_1 drops out of the basic set of variables. Otherwise, return to Step I.

It will be shown that during a major cycle w_1 increases to zero. At this point, a new complementary basic solution is obtained. However, the number of basic variables with negative values is at least one less than at the beginning of the major cycle. Since there are at most p negative basic variables, no more than p major cycles are required to obtain a complementary feasible solution of (22). The proof depends on certain properties of matrices invariant under principal pivoting.

Principal pivot transform of a matrix. Consider the homogeneous system $v = Mu$ where M is a square matrix. Here the variables $v_1, \cdots, v_p$ are basic and expressed in terms of the nonbasic variables $u_1, \cdots, u_p$. Let any subset of the v_i be made nonbasic and the corresponding u_i basic. Relabel the full set of basic variables $\bar{v}$ and the corresponding nonbasic variables $\bar{u}$. Let $\bar{v} = \bar{M}\bar{u}$ express the new basic variables $\bar{v}$ in terms of the nonbasic ones. The matrix $\bar{M}$ is called a *principal pivot transform* of M. Of course, this transformation can be carried out only if the principal submatrix of M corresponding to the set of variables z_i and w_i interchanged is nonsingular, and this will be assumed whenever the term is used.

THEOREM 8 (TUCKER [24]): *If a square matrix M has positive principal minors, so does every principal pivot transform of M.*

The proof of this theorem is easily obtained inductively by exchanging the roles of one complementary pair and evaluating the resulting principal minors in terms of those of M.

THEOREM 9: *If a matrix M is positive definite or positive semidefinite so is every principal pivot transform of M.*

Proof: The original proof given by the authors was along the lines of that for the preceding theorem. P. Wolfe has suggested the following elegant proof. Consider $v = Mu$. After the principal pivot transformation, let $\bar{v} = \bar{M}\bar{u}$, where $\bar{u}$ is the new set of nonbasic variables. We wish to show that $\bar{u}\bar{M}\bar{u} = \bar{u}\bar{v} > 0$ if $uMu = uv > 0$. If M is positive definite, the latter is true if $u \neq 0$, and the former must hold because every pair $(\bar{u}_i, \bar{v}_i)$ is identical with (u_i, v_i) except possibly in reverse order. Hence $\sum_i \bar{u}_i\bar{v}_i = \sum_i u_iv_i > 0$. The proof in the semidefinite case replaces the inequality $>$ by $\geq$.

Validity of the algorithm. The proof given below for $p = 3$ goes through for general p. Consider

$$w_1 = q_1 + m_{11}z_1 + m_{12}z_2 + m_{13}z_3,$$
$$w_2 = q_2 + m_{21}z_1 + m_{22}z_2 + m_{23}z_3,$$
$$w_3 = q_3 + m_{31}z_1 + m_{32}z_2 + m_{33}z_3.$$

Suppose that M has positive principal minors so that the diagonal coefficients are all positive:

$$m_{11} > 0, \qquad m_{22} > 0, \qquad m_{33} > 0.$$

Suppose furthermore that some q_i is negative, say $q_1 < 0$. Then the solution $(w; z) = (q_1, q_2, q_3; 0, 0, 0)$ is complementary, but not feasible because a particular variable, in this case w_1, which we refer to as *distinguished* is negative. We now initiate an almost-complementary path by increasing the complement of the distinguished variable, in this case z_1, which we call the *driving* variable. Adjusting the basic variables, we have

$$(w; z)^1 = (q_1 + m_{11}z_1, q_2 + m_{21}z_1, q_3 + m_{31}z_1; 0, 0, 0).$$

Note that the distinguished variable w_1 increases strictly with the increase of the driving variable z_1 because $m_{11} > 0$. Assuming nondegeneracy, we can increase z_1 by a positive amount before it is blocked either by w_1 reaching zero or by a basic variable that was positive and is now turning negative.

In the former case, for some positive value z_1^* of the driving variable z_1, we have $w_1 = q_1 + m_{11}z_1^* = 0$. The solution

$$(w; z)^2 = (0, q_2 + m_{21}z_1^*, q_3 + m_{31}z_1^*; 0, 0, 0)$$

is complementary and has one less negative component. Pivoting on m_{11}, replaces w_1 by z_1 as a basic variable. By Theorem 8, the matrix $\bar{M}$ in the new canonical system relabeled $\bar{w} = \bar{q} + \bar{M}\bar{z}$ has positive principal minors, allowing the entire major cycle to be repeated.

In the latter case, we have some other basic variable, say $w_2 = q_2 + m_{21}z_1$ blocking when $z_1 = z_1^* > 0$. Then clearly $m_{21} < 0$ and $q_2 > 0$. In this case,

$$(w; z)^2 = (q_1 + m_{11}z_1^*, 0, q_3 + m_{31}z_1^*; z_1^*; 0, 0).$$

THEOREM 10: *If the driving variable is blocked by a basic variable other than its complement, a principal pivot exchanging the blocking variable with its complement will permit the further increase of the driving variable.*

Proof: Pivoting on m_{22} generates the canonical system

$$w_1 \quad = \bar{q}_1 + \bar{m}_{11}z_1 + \bar{m}_{12}w_2 + \bar{m}_{13}z_3,$$

$$z_2 \quad = \bar{q}_2 + \bar{m}_{21}z_1 + \bar{m}_{22}w_2 + \bar{m}_{23}z_3,$$

$$w_3 = \bar{q}_3 + \bar{m}_{31}z_1 + \bar{m}_{32}w_2 + \bar{m}_{33}z_3.$$

The solution $(w; z)^2$ must satisfy the above since it is an equivalent system. Therefore setting $z_1 = z_1^*$, $w_2 = 0$, $z_3 = 0$ yields

$$(w; z)^2 = (q_1 + \bar{m}_{11}z_1^*, 0, q_3 + \bar{m}_{31}z_1^*; z_1^*, 0, 0)$$

i.e. the same almost-complementary solution. Increasing z_1 beyond z_1^* yields

$$(\bar{q}_1 + \bar{m}_{11}z_1, 0, \bar{q}_3 + m_{31}z_1; z_1, 0, 0)$$

which is also almost-complementary. The sign of $\bar{m}_{21}$ is the reverse of m_{21}, since $\bar{m}_{21} = -m_{21}/m_{22} > 0$. *Hence z_2 increases with increasing $z_1 > z_1^*$*; i.e. the new basic variable replacing w_2 is not blocking. Since M has positive principal minors, $\bar{m}_{11} > 0$. *Hence w_1 continues to increase with increasing $z_1 > z_1^*$.*

THEOREM 11: *The number of iterations within a major cycle is finite.*

Proof: There are only finitely many possible bases. No basis can be repeated with a larger value of z_1. To see this, suppose it did for $z_1^{**} > z_1^*$. This would imply that some component of the solution turns negative at $z_1 = z_1^*$ and yet is nonnegative when $z_1 = z_1^{**}$. Since the value of a component is linear in z_1 we have a contradiction.

Paraphrase of the principal pivoting method. Along the almost-complementary path there is only one degree of freedom. In the proof of the validity of the algorithm, z_1 was increasing and z_2 was shown to increase. The same class of solutions can be generated by regarding z_2 as the driving variable and the other variables as adjusting. Hence within each major cycle, the same almost-complementary path can be generated as follows. The first edge is obtained by using the complement of the distinguished variable as the driving variable. As soon as the driving variable is blocked, the following steps are iterated:

(a) replace the blocking variable by the driving variable and terminate the major cycle if the blocking variable is distinguished; if the blocking variable is not distinguished;

(b) let the complement of the blocking variable be the new driving variable and increase it until a new blocking variable is identified; return to (a).

The paraphrase form is used in practice.

THEOREM 12. *The principal pivoting method terminates in a solution of* (1), (2) *if M has positive principal minors (and, in particular, if M is positive definite).*

Proof: We have shown that the completion of a major cycle occurs in a finite number of steps, and each one reduces the total number of variables with negative values. Hence in a finite number of steps, this total is reduced to zero and a solution of the fundamental problem (1), (2) is obtained. Since a positive definite matrix has positive principal minors, the method applies to such matrices.

As indicated earlier, the positive semidefinite case can be handled by using the paraphrase form of the algorithm with a minor modification. The reader will find details in [4].

REFERENCES

1. Cottle, R. W., "Symmetric dual quadratic programs," Quart. Appl. Math., **21** (1963), 237–243.
2. ———, "Nonlinear programs with positively bounded Jacobians," J. SIAM Appl. Math., **14** (1966), 147–158.
3. Dantzig, G. B., *Linear Programming and Extensions*, Princeton Univ. Press, Princeton, N. J., 1963.
4. Dantzig, G. B., and R. W. Cottle, *Positive (Semi) Definite Programming*, ORC 63-18 (RR), May 1963, Operations Research Center, University of California, Berkeley. Revised in *Nonlinear Programming*, edited by J. Abadie, Amsterdam: North-Holland, 1967, pp. 55–73.
5. Dorn, W. S., "Duality in quadratic programming," Quart. Appl. Math., **18** (1960), 155–162.
6. Du Val, P., "The unloading problem for plane curves," Amer. J. Math., **62** (1940), 307–311.

7. Farkas, J., "Theorie der einfachen Ungleichungen," J. Reine Angew. Math., 124 (1902), 1–27.

8. Fiedler, M., and V. Ptak, "On matrices with non-positive off-diagonal elements and positive principal minors," Czech. Math. Journal, 12 (1962), 382–400.

9. ———, "Some generalizations of positive definiteness and monotonicity," Numerische Math., 9 (1966), 163–172.

10. Gale, D., and H. Nikaido, "The Jacobian matrix and global univalence of mappings," Math. Ann., 159 (1965), 81–93.

11. Goldman, A. J., "Resolution and separation theorems for polyhedral convex sets" in *Linear Inequalities and Related Systems*, edited by H. W. Kuhn and A. W. Tucker, Princeton, N. J.: Princeton Univ. Press, 1956.

12. Hall, Jr., M., and M. Newman, "Copositive and completely positive quadratic forms," Proc. Camb. Philos. Soc., 59 (1963), 329–339.

13. Kilmister, C. W., and J. E. Reeve, *Rational Mechanics*, New York: American Elsevier, 1966, §5.4.

14. Kuhn, H. W., and A. W. Tucker, "Nonlinear programming" in *Second Berkeley Symposium on Mathematical Statistics and Probability*, edited by J. Neyman, Berkeley, Calif.: Univ. of California Press, 1951.

15. Lemke, C. E., and J. T. Howson, Jr., "Equilibrium points of bimatrix games," J. Soc. Indust. Appl. Math., 12 (1964), 413–423.

16. Lemke, C. E., "Bimatrix equilibrium points and mathematical programming," Management Sci., 11 (1965), 681–689.

17. ———, Private communication.

18. Motzkin, T. S., "Copositive quadratic forms," Nat. Bur. Standards Report No. 1818, 1952, pp. 11–12.

19. Nash, J. F., "Noncooperative games," Ann. Math., 54 (1951), 286–295.

20. von Neumann, J., "Discussion of a maximum problem" in *Collected Works*. Vol. VI, edited by A. H. Taub, New York: Pergamon Press, 1963.

21. van de Panne, C., and A. Whinston, "A comparison of two methods for quadratic programming," Operations Res., 14 (1966), 422–441.

22. Parsons, T. D., "A combinatorial approach to convex quadratic programming," Doctoral Dissertation, Department of Mathematics, Princeton University, 1966.

23. Tucker, A. W., "A combinatorial equivalence of matrices," *Proc. Sympos. Appl. Math.*, Vol. 10, edited by R. Bellman and M. Hall, Amer. Math Soc., Providence, R. I., 1960.

24. ———, "Principal pivotal transforms of square matrices," SIAM Review, 5 (1963), 305.

25. ———, *Pivotal Algebra*, Lecture notes (by T. D. Parsons), Department of Mathematics, Princeton University, 1965.

26. Wolfe, P., "The simplex method for quadratic programming," Econometrica, 27 (1959), 382–398.

"STEINER'S" PROBLEM REVISITED

*Harold W. Kuhn**

1. INTRODUCTION

The subject of this paper is one of the oldest optimization problems in mathematics. Each time I look at it again, I discover new literature related to it. Since most of this literature consists of inadvertent rediscovery of previous results (of which I have been guilty, myself), why add another paper? Two reasons seem to make this worthwhile. First, the problem is a beautiful one that deserves exposition of its several aspects in a modern style. Secondly, to my knowledge, there is no single paper that includes both a treatment of the problem as a nonlinear program with its "dual" and a correct description of the "natural" algorithm for the problem. These are the objectives of this paper. As such, it contains material that appeared originally in [3], [4], [5], and [6], organized to give a view of the whole.

* This paper was written while the author was Science Faculty Fellow of the National Science Foundation at the London School of Economics.

The history of the problem (and the reason for the quotes around Steiner in the title) is sketched in Section 2. Of special interest in this account is the result of Fasbender, which provides an example over a hundred years old of a duality between nonlinear programs. Of more practical interest is the algorithm of Weiszfeld which, although discovered thirty-five years ago, is extraordinarily modern in its approach.

The basic properties of the problem are established in Section 3 as preparation for the other sections. The generalization of Fasbender's geometric duality is given in Section 4 and a convergence proof for Weiszfeld's algorithm is the subject of Section 5.

2. A HISTORICAL SKETCH

Our story starts with a problem rather casually posed by Fermat early in the 17th century. At the end of a celebrated essay on maxima and minima, in which he presented pre-calculus rules for finding tangents to a variety of curves, he threw out the challenge: "Let he who does not approve of my method attempt the solution of the following problem: Given three points in the plane, find a fourth point such that the sum of its distances to the three given points is a minimum!" The problem seems to have travelled to Italy with Mersenne; it is known that before 1640 Torricelli had proposed a geometric solution to the problem. He asserted that the circles circumscribing the equilateral triangles constructed on the sides of and outside the given triangle intersect in the point that is sought. This point is called the *Torricelli point*. In Cavalieri's *Exercitationes Geometricae* of 1647, it is shown that the sides of the given triangle subtend angles of 120° from the Torricelli point. Furthermore, Simpson asserted and proved in his *Doctrine and Application of Fluxions* (London, 1750) that the three lines joining the outside vertices of the equilateral triangles defined above to the opposite vertices of the given triangle intersect in the Torricelli point. These three lines are called *Simpson lines*. There are, of course, exceptional cases and these have continued to complicate treatments whether by geometric or by analytic methods. Apparently, the first result that recognizes these cases explicitly is due to F. Heinen who proved in 1834 that, for a triangle in which one angle is greater than or equal to 120°, the vertex of this angle is the minimizing point; he also proved that the lengths of the three Simpson lines are equal to the minimum sum of distances.

The duality we shall study below originates with a result of Fasbender [2]* published in 1846. He proved that the perpendiculars to the Simpson lines through the three given points are the sides of the *largest* equilateral triangle circumscribing these points and that the altitude of this triangle equals the minimum sum of distances. We shall see how this theorem is a remarkable instance of nonlinear duality.

The interest of economists in the problem appears to stem from the pioneering work *Über den Standort des Industrien* of Alfred Weber (Tübingen, 1909) who posed the problem of minimizing the *weighted* sum of distances from given points. (Actually, as a purely mathematical problem, this appears as an exercise in Simpson's "Fluxions".) In view of its potential application to practical situations, it is natural that recent interest in the problem has centered on computational methods. However, here again, the "natural" algorithm was discovered even earlier as pure mathematics by Weiszfeld in 1937 [7] and rediscovered several times since then. Its properties are established in this paper.

No historical sketch of this problem would be complete without mention of the fact that the Fermat Problem has been widely popularized by Courant and Robbins [1] under the name of the "Steiner Problem". Although this gifted geometer of the 19th century can be counted among the dozens of mathematicians who have written on the subject, he does not seem to have contributed anything new, either to its formulation or to its solution. As for the statements of Courant and Robbins that the generalization of the problem to more than three points is sterile, their answer is to be found in the new applications and understanding that have come through this "sterile" extension.

---

* Numbers in square brackets [ ] refer to the references at the end of the paper; for this historical section, the reader is referred to the article by M. Zacharias in the *Encyklopädie der Mathematischen Wissenschaften*, III AB9.

3. BASIC PROPERTIES

Although the simpler properties of the problem have been amply discussed in the literature, for the sake of completeness and to establish notation, we shall restate the problem and derive some basic results here.

Let there be given m points $A_i = (a_{i1}, \cdots, a_{in})$, called *vertices*, and m positive numbers w_i, called *weights*. Furthermore, for $P = (x_1, \cdots, x_n)$, let

$$d_i(P) = \sqrt{\Sigma_j(x_j - a_{ij})^2},$$

the Euclidean distance from P to A_i, for $i = 1, \cdots, m$.

GENERAL FERMAT PROBLEM. Find a point that minimizes $f(P) = \Sigma_i w_i d_i(P)$.

By the triangle inequality for Euclidean distance, $d_i(P)$ is a convex function of P. Indeed, it is a strictly convex function of P except on the halflines ending at A_i; on each of these halflines it is a linear function. A positive multiple of a (strictly) convex function is (strictly) convex and the sum of (strictly) convex functions is (strictly) convex. Hence, we have proved the following results:

(3.1) *If the vertices A_i are not collinear then $f(P)$ is a strictly convex function of P. If the points are collinear, then $f(P)$ is piecewise linear and convex on the line through them and strictly convex elsewhere.*

(3.2) *If the vertices A_i are not collinear then $f(P)$ has a* **unique** *minimum $P = M$.*

We shall only consider non-collinear problems; the problems excluded by this restriction are clearly trivial. For the rest of the paper we shall use M to denote the unique solution to the problem.

Motives both mathematical and physical (deriving from a string and weight model of the problem introduced by G. Pick as early as 1909) suggest the introduction of the negative of the gradient of f (i.e., the *resultant* of the forces in the strings).

To this end, let

$$R(P) = \Sigma_i \frac{w_i}{d_i(P)} (A_i - P) \quad \text{if} \quad P \neq A_i \quad \text{for all } i.$$

Obviously, R is not defined at any vertex A_i. However, by physical analogy, set

$$R_k = \Sigma_{i \neq k} \frac{w_i}{d_i(A_k)} (A_i - A_k) \qquad \text{for} \quad k = 1, \cdots, m$$

and extend the definition of R by setting

$$R(A_k) = \max(|R_k| - w_k, 0) \frac{R_k}{|R_k|} \qquad \text{for} \quad k = 1, \cdots, m.$$

(Here, as elsewhere in this paper, $|A|$ denotes the length of the vector A.) In the expression for $R(A_k)$, the length of R_k is compared with w_k. If $w_k \geqq |R_k|$ then $R(A_k) = 0$; otherwise, a "resultant" of magnitude $|R_k| - w_k$ is defined in the direction of R_k.

(3.3) *The point $P = M$ if and only if $R(P) = 0$.*

Proof: If P is not a vertex, then the convexity and differentiability of f imply that the first-order conditions $R(P) = 0$ are both necessary and sufficient for a minimum.

If $P = A_k$ then consider a change from A_k to $A_k + tZ$ for $|Z| = 1$. Then direct calculation yields:

$$\frac{d}{dt} f(A_k + tZ) = w_k - R_k \cdot Z \qquad \text{for} \quad t = 0,$$

and hence the direction of greatest decrease of f from A_k is $Z = R_k/|R_k|$. (Here, $A \cdot B$ denotes the inner product of A and B and A^2 will be used as an abbreviation for $A \cdot A$.) Clearly, A_k is a local minimum if and only if

$$w_k - \frac{R_k^2}{|R_k|} \geqq 0,$$

which is the same as $R(A_k) = 0$. Again, the convexity of f implies that $R(A_k) = 0$ is both necessary and sufficient for A_k to be a global minimum.   Q.E.D.

(3.4) *The point M is in the convex hull of the vertices A_i.*

Proof: If M is a vertex, then it is trivially in the convex hull.

Otherwise, the condition $R(M) = 0$ yields the equations:

$$M = \Sigma_i \frac{w_i}{d_i(M)} A_i / \Sigma_i \frac{w_i}{d_i(M)} .$$

Thus, M is a weighted sum of the vertices with positive weights that sum to one. Q.E.D.

(3.5) *If M is not a vertex then M is not on the boundary of the convex hull of the vertices A_i.*

Proof: This is simply a consequence of the fact that the condition $R(M) = 0$ yields an expression for M with *positive* weights.

Proposition (3.4) can be sharpened to a result which asserts that every point B outside the convex hull is at least as far away from each of the A_i than some point A (which will depend on B) in the convex hull. This result, which was first proved in [4], provides a curious and novel characterization of the convex hull of a finite set of points. We shall say that A *dominates* B with respect to the points A_i if $A \neq B$ and $d_i(A) \leqq d_i(B)$ for $i = 1, \cdots, m$. A point is said to be *admissible* if it is not dominated by any other point.

(3.6) *A point A is admissible if and only if it is in the convex hull of the points A_i.*

Proof: We first prove that every point B not in the convex hull is dominated by some point A in the convex hull. If this is not the case, then for all A in the convex hull,

$$d_i(A) > d_i(B) \qquad \text{for some } i.$$

This means that the functions $\varphi_i(A) = d_i(A) - d_i(B)$ form a finite family of continuous functions on a compact convex set which is such that

$$\max \varphi_i(A) > 0 \qquad \text{for all } A.$$

Hence, by Ville's Lemma, there exist $w_i \geqq 0$ with $\Sigma_i w_i = 1$ such that $\Sigma_i w_i d_i(A) \geqq \Sigma_i w_i d_i(B)$ for all A in the convex hull. However, this means that B solves the General Fermat Problem defined on those points A_i with positive weights w_i and contradicts (3.6).

We must now show that every point of the convex hull is admissible. Clearly, all of the vertices A_i are admissible. Let A and B be two points in the convex hull of the A_i and assume C is an *inadmissible* point on the segment $\overline{AB}$ joining A to B. If C is dominated by D then all of the vertices A_i lie on the same side of the plane perpendicular bisector of $\overline{CD}$ as does D. Then A and B also lie in this halfspace since they are in the convex hull. This means C does as well, which is a contradiction. To recapitulate, we have proved that the A_i are admissible and that the segment joining any two points in the convex hull consists of admissible points. Hence all points in the convex hull are admissible.   Q.E.D.

4. FASBENDER'S DUALITY

To state and prove the remarkable result discovered by Fasbender in 1846, let us return to the original Fermat Problem: Given three points A_1, A_2, and A_3 in the plane, find a fourth point that minimizes

$$d_1(P) + d_2(P) + d_3(P),$$

where $d_i(P)$ is the distance from P to A_i for $i = 1, 2, 3$. For the moment, we shall restrict ourselves (as did Fasbender) to the special case in which the problem has its solution $P = M$ in the interior of the triangle $\Delta A_1 A_2 A_3$.

(4.1) *Let Q be any point interior to an equilateral triangle $\Delta B_1 B_2 B_3$ with altitude h. Let h_i be the length of the perpendicular from Q to the side opposite B_i* (see Figure 4.1) *for $i = 1, 2, 3$. Then $h_1 + h_2 + h_3 = h$.*

Proof: Area $\Delta B_1 B_2 B_3$ = Area $\Delta Q B_2 B_3$ + Area $\Delta Q B_3 B_1$ + Area $\Delta Q B_1 B_2$. Hence

$$\frac{h^2}{\sqrt{3}} = \frac{h}{\sqrt{3}} (h_1 + h_2 + h_3)$$

and the result is proved.   Q.E.D.

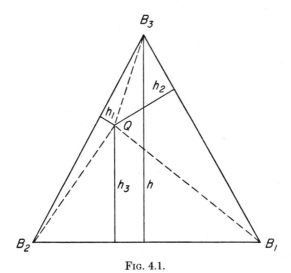

FIG. 4.1.

(4.2) *Given three points A_1, A_2, and A_3 in the plane and any equilateral triangle with altitude h circumscribing $\Delta A_1A_2A_3$, for any point P interior to the equilateral triangle,*

$$h \leqq d_1(P) + d_2(P) + d_3(P) = f(P).$$

Proof: This result follows immediately from (4.1) and the inequalities $h_i \leqq d_i(P)$, illustrated in Figure 4.2, where each h_i is the leg of a right triangle with hypotenuse $d_i(P)$.

FASBENDER'S THEOREM: *If the solution M to the Fermat Problem is not a vertex then $f(M)$ is both the minimum distance sum and the maximum altitude of a circumscribing equilateral triangle. This triangle has its sides perpendicular to the segments $\overline{MA_i}$ for $i = 1, 2, 3$.*

Proof: We have shown that $h \leqq f(P)$ in (4.2). Equality between these two quantities is sufficient to prove that we have the circumscribing triangle with maximum altitude and the point $P = M$ with the minimum distance sum. However, the proof of (4.2) shows that we have equality if and only if the segments $\overline{PA_i}$ are perpendicular to the sides of the equilateral triangle, that is, if and only if the sides of $\Delta A_1A_2A_2$ subtend angles of $120°$ at P.

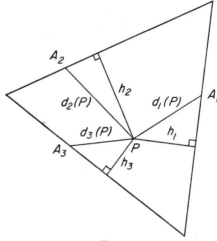

However, this is exactly the condition for $P = M$, a non-vertex solution to the Fermat Problem.

This remarkable theorem contains echos of the properties of linear programming in a nonlinear situation. Although several theories of duality for nonlinear programming have been proposed, in most instances the dual program is difficult to give explicitly. Much more serious, in almost every case, its mere statement requires a solution to the original or primal problem. Therefore, it comes as a surprise that the General Fermat Problem possesses a dual without either of these drawbacks. Although this duality was discovered prior to my knowing of the Fasbender result, it will be seen that it generalizes his theorem in a precise fashion. (After the dual was found in 1963, C. Witzgall informed me that he and R. T. Rockafellar had discovered the same result.)

GENERAL FERMAT DUAL PROBLEM. Find m vectors U_i in n-space that maximize $g(U_1, \cdots, U_m) = \Sigma_i U_i \cdot A_i$ subject to $\Sigma_i U_i = 0$ and $|U_i| \leq w_i$ for $i = 1, \cdots, m$.

It should be noted immediately that this dual problem is constructed merely from the data of the original problem and does not require its solution.

(4.3) *For any P and any feasible U_i,*

$$g(U_1, \cdots, U_m) \leqq f(P).$$

Proof:

$$g(U_1, \cdots, U_m) = \Sigma_i U_i \cdot A_i - (\Sigma_i U_i) \cdot P = \Sigma_i U_i \cdot (A_i - P).$$

However,

$$U_i \cdot (A_i - P) \leqq |U_i| d_i(P)$$

by the Schwartz inequality, and hence

$$g(U_1, \cdots, U_m) = \Sigma_i U_i \cdot (A_i - P) \leqq \Sigma_i |U_i| d_i(P) \leqq \Sigma_i w_i d_i(P)$$
$$= f(P).$$

Proposition (4.3) implies that equality of $f(P)$ and $g(U_1, \cdots, U_m)$ for any P and feasible U_i is a condition *sufficient* to establish the optimality of both P and the U_i. We now turn to necessity.

(4.4) *If M solves the General Fermat Problem, then there exist feasible U_i such that*

$$g(U_1, \cdots, U_m) = f(M).$$

Proof: We distinguish two cases: (1) M is not a vertex; (2) $M = A_k$ for some k.

Case 1. In this case, set

$$U_i = \frac{w_i}{d_i(M)} (A_i - M) \qquad \text{for} \quad i = 1, \cdots, m.$$

Then, by (3.3) and the definition of R,

$$\Sigma_i U_i = R(M) = 0,$$

and

$$|U_i| = \frac{w_i}{d_i(M)} |A_i - M| = w_i \qquad \text{for} \quad i = 1, \cdots, m.$$

Therefore the U_i are feasible and both of the inequalities in the proof of (4.3) become equations, the first because $U_i \cdot (A_i - M) = |U_i| d_i(M)$ and the second because $|U_i| = w_i$ for $i = 1, \cdots, m$.

Case 2. Suppose $M = A_k$; set

$$U_i = \frac{w_i}{d_i(A_k)} (A_i - A_k) \qquad \text{for} \quad i \neq k,$$

and define

$$U_k = -\Sigma_{i \neq k} U_i.$$

Then $\Sigma_i U_i = 0$ by definition, while $| U_i | = w_i$ for $i \neq k$ as before. Finally, $U_k = -R_k$ and $R(X_k) = 0$ imply $w_k \geqq | R_k | = | U_k |$. Therefore, the U_i defined in this manner are feasible.

Again, both of the inequalities in the proof of (4.3) become equations. The only change in the previous argument is in the kth term. Here

$$U_k \cdot (A_k - M) = | U_k | d_k(M) = w_k d_k(M)$$

since all three terms are zero.

Proposition (4.4) implies that any solution to the General Fermat Problem provides a solution to the dual in a straightforward and explicit manner.

(4.5) *Given any feasible U_i that solve the dual problem, there exists a P such that*

$$g(U_1, \cdots, U_m) = f(P),$$

and hence $P = M$.

Proof: Given feasible U_i that solve the dual problem, we may formulate the Kuhn-Tucker necessary conditions for this nonlinear program. The appropriate Lagrangian function is:

$$\Sigma_i U_i \cdot A_i - P \cdot \Sigma_i U_i - \tfrac{1}{2} \Sigma_i t_i (w_i^2 - U_i^2),$$

where the components of $P = (x_1, \cdots, x_n)$ and the t_i are multipliers. Then there exist multipliers such that

$$A_i - P - t_i U_i = 0$$
$$\Sigma_i U_i = 0$$
$$w_i^2 - U_i^2 \geqq 0$$
$$t_i \geqq 0$$
$$t_i (w_i^2 - U_i^2) = 0$$

for $i = 1, \cdots, m$. Again, we must treat two separate cases.

Case 1. All $t_i > 0$. Then $|\,U_i\,| = w_i$ and $t_i = d_i(P)/w_i$ for $i = 1, \cdots, m$. Since

$$U_i = \frac{w_i}{d_i(P)} (A_i - P),$$

we again have equality of the objective functions and $P = M$.

Case 2. Some $t_i = 0$, say, for $i = k$. Then $P = A_k$ and, since the points A_i are distinct, we must have $t_i > 0$ for $i \neq k$. Hence $|\,U_i\,| = w_i$ and

$$U_i = \frac{w_i}{d_i(A_k)} (A_i - A_k) \qquad \text{for} \quad i \neq k$$

as before. Equality between the two objective functions follows exactly as in Case 2 of (4.4).

Proposition (4.5) implies that any solution to the dual provides a solution to the General Fermat Problem in a direct and explicit manner. If all $|\,U_i\,| = w_i$, then the line through A_i with direction U_i passes through M. To find M, two lines will suffice. Otherwise, at most one U_i satisfies $|\,U_i\,| < w_i$, say, for $i = k$. Then $M = A_k$ is the desired solution.

This completes the proofs of the properties which parallel in a remarkable way the duality of linear programming. It only remains to show that the dual problem generalizes in a precise fashion the "geometric duality" discovered by Fasbender. This is easily verified using Figure 4.2. Let U_i be a unit vector from P and perpendicular to the side of the equilateral triangle circumscribing $\Delta A_1 A_2 A_3$ which contains A_i for $i = 1, 2, 3$. Clearly the U_i so defined are feasible for the dual associated with the simple Fermat Problem. Furthermore, $h_i = U_i \cdot (A_i - P)$ for $i = 1, 2, 3$, and hence $h = \Sigma_i h_i = \Sigma_i U_i \cdot (A_i - P)$. Since $\Sigma_i U_i = 0$, we have $h = \Sigma_i U_i \cdot A_i$ and the altitude of the circumscribing triangle equals the objective function associated with the U_i. Conversely, any feasible U_i determine a circumscribing equilateral triangle, with the same equality holding. Thus, Fasbender's maximum problem is a special case of the dual to the General Fermat Problem.

The theory of subdifferentials and directional derivatives, which has been shown by Rockafellar and Moreau to be a key technique in the general theory of duality, sheds considerable light on the pair of dual programs. These connections have been sketched in [5], where the dual to the General Fermat Problem was first published.

5. WEISZFELD'S ALGORITHM

The equation

$$M = \Sigma_i \frac{w_i}{d_i(M)} A_i / \Sigma_i \frac{w_i}{d_i(M)} ,$$

used in the proof of (3.4) suggests quite naturally a method of successive approximation. For P *not* a vertex, define

$$T: P \to T(P) = \Sigma_i \frac{w_i}{d_i(P)} A_i / \Sigma_i \frac{w_i}{d_i(P)} .$$

For the sake of continuity, set

$$T(A_i) = A_i \qquad \text{for} \quad i = 1, \cdots, m.$$

Then, as an immediate consequence of the definition of T, the equation above, and (3.3):

(5.1) *If $P = M$ then $T(P) = P$. If P is not a vertex and $T(P) = P$ then $P = M$.*

In effect, the algorithm proposed is merely a simple attempt to solve the first-order conditions $R(P)$ iteratively. It seems to have been discovered by E. Weiszfeld in 1937 [7] who asserted that, for any P_0 that is not a vertex, the sequence $P_r = T^r(P_0)$ converges to M. In this section, we shall investigate the properties of T and prove a corrected statement of this theorem.

First, note that the algorithm proposed is a "long-step" gradient method. Indeed, recalling that $-R(P)$ is the gradient of f whenever it exists, direct calculation yields

$$T(P) = P + h(P)R(P)$$

where

$$h(P) = \Pi_i d_i(P)/\Sigma_k(w_k\Pi_{i\neq k}d_i(P))$$

for *all* points P. Thus, the algorithm follows the direction of the resultant with precalculated length of step $h(P) \mid R(P) \mid$. Apart from the vertices, which are all left fixed by T, one difficulty with such methods is that they may "overshoot". The following result (first proved in [7]) shows that this is not the case.

(5.2) *If* $T(P) \neq P$ *then* $f(T(P)) < f(P)$.

Proof: Since $T(P) \neq P$, P is not a vertex and

$$T(P) = \Sigma_i \frac{w_i}{d_i(P)} A_i/\Sigma_i \frac{w_i}{d_i(P)}.$$

This says that $T(P)$ is the center of gravity of weights $w_i/d_i(P)$ placed at the vertices A_i. Hence, by elementary calculus, $T(P)$ is the unique minimum of the strictly convex function

$$g(Q) = \Sigma_i \frac{w_i}{d_i(P)} d_i^2(Q).$$

Since $P \neq T(P)$,

$$g(T(P)) = \Sigma_i \frac{w_i}{d_i(P)} d_i^2(T(P)) < g(P) = \Sigma_i \frac{w_i}{d_i(P)} d_i^2(P) = f(P).$$

On the other hand,

$$g(T(P)) = \Sigma_i \frac{w_i}{d_i(P)} [d_i(P) + (d_i(T(P)) - d_i(P))]^2$$

$$= f(P) + 2(f(T(P)) - f(P))$$

$$+ \Sigma_i \frac{w_i}{d_i(P)} [d_i(T(P)) - d_i(P)]^2.$$

Combining these results,

$$2f(T(P)) + \Sigma_i \frac{w_i}{d_i(P)} [d_i(T(P)) - d_i(P)]^2 < 2f(P)$$

and the assertion $f(T(P)) < f(P)$ is proved.   Q.E.D.

A second possible difficulty with the algorithm is that the sequence of approximations might remain in the neighborhood of a non-optimal vertex. The following result shows that this cannot happen. Informally, it says that there is a neighborhood of each non-optimal vertex such that, if the approximation sequence enters it, then it is eventually "kicked out" by T.

(5.3) *Suppose $A_k \neq M$. Then there exists $\delta > 0$ such that $0 < d_k(P) \leq \delta$ implies $d_k(T^s(P)) > \delta$ and $d_k(T^{s-1}(P)) \leq \delta$ for some positive integer s.*

Proof:

$$T(P) - A_k = P + h(P)R(P) - A_k$$

$$= h(P) \, \Sigma_{i \neq k} \frac{w_i}{d_i(P)} \, (A_i - P)$$

$$+ \left(\frac{h(P) w_k}{d_k(P)} - 1 \right) (A_k - P).$$

Since $A_k \neq M$, we have $| \, \Sigma_{i \neq k}(w_i/d_i(A_k))(A_i - A_k) \, | > w_k$. Hence, there exist $\delta' > 0$ and $\epsilon > 0$ such that

$$\left| \Sigma_{i \neq k} \frac{w_i}{d_i(P)} \, (A_i - P) \right| \geq (1 + 2\epsilon) w_k \quad \text{for} \quad d_k(P) \leq \delta'.$$

By the definition of h, we have $\lim_{P \to A_k} h(P) w_k / d_k(P) = 1$. Hence, there exists $\delta'' > 0$ such that

$$\left| \frac{h(P) w_k}{d_k(P)} - 1 \right| < \frac{\epsilon}{2(1 + \epsilon)} \quad \text{for} \quad 0 < d_k(P) \leq \delta''.$$

Set $\delta = \min(\delta', \delta'')$. For $0 < d_k(P) \leq \delta$, we have

$$d_k(T(P)) > h(P)(1 + 2\epsilon) w_k - \frac{\epsilon}{2(1 + \epsilon)} \, d_k(P)$$

$$> \left(1 - \frac{\epsilon}{2(1 + \epsilon)} \, (1 + 2\epsilon) d_k(P) - \frac{\epsilon}{2(1 + \epsilon)} \, d_k(P) \right)$$

$$= (1 + \epsilon) d_k(P).$$

Since $d_k(P) > 0$, $(1 + \epsilon)^t d_k(P) > \delta$ for some positive integer t and hence $d_k(T^s(P)) > \delta$ for some positive integer s with $d_k(T^{s-1}(P)) \leqq \delta$.

The following result (first proved in [7]), which could be used to derive (5.3), describes the behavior of T near all vertices, optimal or not.

$$(5.4) \quad \lim_{P \to A_k} \frac{d_k(T(P))}{d_k(P)} = \frac{|R_k|}{w_k} \quad \text{for} \quad k = 1, \cdots, m.$$

Proof: For P not a vertex,

$$T(P) = \Sigma_i \frac{w_i}{d_i(P)} A_i / \Sigma_i \frac{w_i}{d_i(P)}$$

$$= \frac{\Sigma_{i \neq k} \dfrac{w_i}{d_i(P)} (A_i - A_k) + A_k \Sigma_i \dfrac{w_i}{d_i(P)}}{\Sigma_i \dfrac{w_i}{d_i(P)}}.$$

Hence

$$T(P) - A_k = \Sigma_{i \neq k} \frac{w_i}{d_i(P)} (A_i - A_k) / \Sigma_i \frac{w_i}{d_i(P)}$$

and

$$\frac{1}{d_k(P)} (T(P) - A_k) = \frac{\Sigma_{i \neq k} \dfrac{w_i}{d_i(P)} (A_i - A_k)}{w_k \left(1 + \dfrac{d_k(P)}{w_k} \Sigma_{i \neq k} \dfrac{w_i}{d_i(P)}\right)}.$$

Taking the limits of the lengths of both sides,

$$\lim_{P \to A_k} \frac{d_k(T(P))}{d_k(P)} = \frac{|R_k|}{w_k}. \qquad \text{Q.E.D.}$$

CONVERGENCE THEOREM: *Given any P_0, define $P_r = T^r(P_0)$ for $r = 1, 2, \cdots$. If no P_r is a vertex then $\lim_{r\to\infty} P_r = M$.*

Proof: With the possible exception of P_0, the sequence P_r lies in the convex hull of the vertices, a compact set. Hence, by the Bolzano-Weierstrass Theorem, there exists at least one point P and a subsequence P_{r_l} such that $\lim_{l\to\infty} P_{r_l} = P$. To prove the theorem we must verify that $P = M$ in all cases.

If $P_{r+1} = T(P_r) = P_r$ for some r, then the sequence repeats from that point and $P = P_r$. Since P_r is not a vertex, $P = M$ by (5.1).

Otherwise, by (5.2),

$$f(P_0) > f(P_1) > \cdots > f(P_r) > \cdots > f(M).$$

Hence $\lim_{r\to\infty}(f(P_{r_l}) - f(T(P_{r_l}))) = 0$. Since the continuity of T implies $\lim_{l\to\infty} T(P_{r_l}) = T(P)$, we have

$$f(P) - f(T(P)) = 0.$$

Therefore, by (5.2), $P = T(P)$. If P is not a vertex then $P = M$ by (5.1). In any event, P lies in the finite set of isolated points $\{A_1, \cdots, A_m, M\}$, where M may be a vertex.

The only case that remains is $P = A_k$ for some k. If $A_k \neq M$, we first isolate A_k from the other vertices (and M if it is not a vertex) by a δ-neighborhood that satisfies (5.3). Then it is clear that we can choose our subsequence $P_{r_l} \to A_k$ so that $d_k(T(P_{r_l})) > \delta$ for all l. This means that the ratio $d_k(T(P_{r_l}))/d_k(P_{r_l})$ is unbounded. However this contradicts (5.4). Hence $A_k = M$ and the theorem is proved.   Q.E.D.

The error in Weiszfeld's statement ([7], p. 356) consists in ignoring the possibility that even if P_0 is chosen distinct from all vertices, some $P_r = T^r(P_0)$ may be a vertex. This may invalidate his arguments (pp. 362–363) where several quotients are then undefined. The following example shows that this is a real possibility and is a counterexample to Weiszfeld's theorem. (A different counterexample has been provided in [6].)

COUNTEREXAMPLE. Consider the four vertices in the plane graphed in Figure 5.1; the associated weights are shown in the

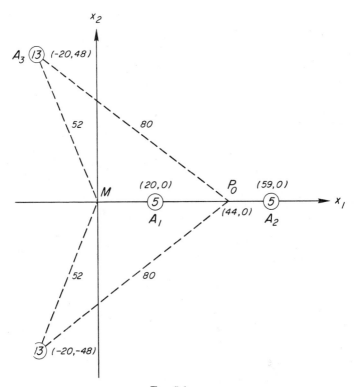

Fig. 5.1

circles at each of the vertices. By direct calculation, the resultant vanishes at the origin and hence $M = (0, 0)$. For the point $P_0 = (44, 0)$, let $T(P_0) = (x_1', x_2')$. Then $d_1(P_0) = 24$, $d_2(P_0) = 15$, $d_3(P_0) = d_4(P_0) = 80$ and hence

$$x_1' = \frac{\dfrac{5}{24}(20) + \dfrac{5}{15}(59) + \dfrac{13}{80}(-20) + \dfrac{13}{80}(-20)}{\dfrac{5}{24} + \dfrac{5}{15} + \dfrac{13}{80} + \dfrac{13}{80}} = 20.$$

By symmetry, or a similar calculation, $x_2' = 0$. Therefore we have

$T(P_0) = A_1$, which is not optimal. If one has the bad luck to start the algorithm from P_0, then $P_1 = A_1$ and the sequence repeats from that point. Thus, the example shows that the sequence P_r need not converge to M even if P_0 is not a vertex.

Of course, this is a very unlikely event. The following result expresses this precisely:

(5.5) *For all but a denumerable number of P_0, $P_r = T^r(P_0)$ converges to M.*

Proof: The Convergence Theorem establishes that, if no P_r is a vertex, then P_r converges to M. To invert T from any vertex A_i, we must solve algebraic equations. Thus we obtain a finite number of P_0 such that $T(P_0) = A_i$. Hence, for a fixed positive r,

$$\{P_0 \mid T^r(P_0) = A_i \text{ for some } i = 1, \cdots, m\}$$

is finite. Finally,

$$\{P_0 \mid T^r(P_0) = A_i \text{ for some } i \text{ and } r\}$$

is denumerable. Q.E.D.

REFERENCES

1. Courant, R. and Robbins, H., *What is Mathematics?*, New York: Oxford University Press, 1941.
2. Fasbender, E., "Über die gleichseitigen Dreiecke, welche um ein gegebenes Dreieck gelegt werden können", *J. f. Math.*, **30** (1846), 230–1.
3. Kuhn, H. W. and Kuenne, R. E., "An efficient algorithm for the numerical solution of the generalized Weber problem in spatial economics", *J. Regional Sci.*, **4** (1962), 21–33.
4. ———, "Locational problems and mathematical programming", in *Colloquium on the Application of Mathematics to Economics* (Budapest, 1965), 235–42.
5. ———, "On a pair of dual nonlinear programs", in *Methods of Nonlinear Programming* (Amsterdam, 1967), 38–54.
6. ———, "A note on Fermat's problem", *Mathematical Programming*, **4** (1973), 98–107.
7. Weiszfeld, E., "Sur le point pour lequel la somme des distances de n points donnés est minimum", *Tôhoku Math. J.*, **43** (1937), 355–86.

PROPERLY LABELED SIMPLEXES

B. Curtis Eaves

0. ON GHOSTS

I eased through the front door of the allegedly haunted house. Just as a ghost appeared, the front door slammed shut behind me. He spoke, "You are now locked inside our house, but it is your fate that except for this room which has one open door, every other room with a ghost has two open doors." I thought, "Is there a room without a ghost?"

1. INTRODUCTION

We consider here a generalization of Sperner's Lemma; the model is defined, examples are developed, and an algorithmic proof is given that there are points with particular limit properties. Our

This research was supported in part by Army Research Office-Durham Contract DAHC-04-71-C-0041 and NSF Grant GP-34559.

structure is somewhat like the fixed point model but differs in its treatment of continuity. Aside from Sperner's Lemma and Dantzig's simplex method of linear programming, the principal motivation for this work can be found in Lemke and Howson [1], Lemke [2], Scarf [3–5], Cohen [6], Fan [7], Kuhn [8], Hansen and Scarf [9], and [10–13]*.

The words "complete" and "properly labeled" defined below are used in much the same sense that they are in the context of Sperner's Lemma. Let R^n be n-dimensional Euclidean space, let S be an (closed) n-simplex in R^n, and let $l: S \to R^n$ be a function from S to R^n.

DEFINITION: Given $l: S \to R^n$, a subset C of S is defined to be l-complete if the origin of R^n is contained in the convex hull of $l(C)$. An element x of S is defined to be an l-complete point if every neighborhood in S of x is l-complete. □

Note that if $\{x\}$ is an l-complete set, then x is an l-complete point, but since we have not made any continuity assumptions the converse is not necessarily so. By a facet of S is meant the convex hull of any n of the $n + 1$ vertices of S.

DEFINITION: The function $l: S \to R^n$ is defined to be a proper labeling of S if

(a) the vertex set of S is l-complete,
(b) no facet of S is l-complete. □

Our main theorem is:

THEOREM 1: *If $l: S \to R^n$ is a proper labeling, then there is an l-complete point in S.* □

---

* In [1, 2] one finds the introduction of a radically new convergence proof and algorithm for the linear complementarity problem. In [3–9] the scheme was adapted to finite pseudomanifolds and applicability of the method was thus extended to include computation of fixed points. In [10–12] a new understanding behind convergence of the algorithm was developed. This explanation was used in [13] to motivate adaption of the scheme to infinite pseudomanifolds; a serious weakness of the earlier methods was thus overcome. The present paper is self-contained and develops a more pedagogical approach to [13].

We say that a point x in S is the limit of a sequence C_k $k = 1, 2,$ $\cdots$ of subsets of S if for each $\epsilon > 0$ there is a k_ϵ such that for $k \geq k_\epsilon$ every point of C_k is within a distance ϵ of x. Clearly x in S is an l-complete point if and only if it is the limit of a sequence of l-complete sets. In our proof of this theorem we develop an algorithm for computing in the limiting sense an l-complete point. The algorithm operates, as one might expect, by generating an infinite sequence of l-complete sets whose diameters tend to zero. Item (a) in the definition of a proper labeling enables the algorithm to get started and item (b) enables it to continue.

2. NOTATION AND LEMMAS

Let N be the set $\{0, 1, \cdots, n\}$ of integers; for $I \subseteq N$ let $N \sim I$ denote the complement of I in N. Given $v = (v^1, \cdots, v^n) \in R^n$ and $u = (u^1, \cdots, u^n) \in R^n$ we denote by $v \cdot u$ the inner product $\Sigma\, v^i u^i$. Let $\| v \|$ be the Euclidean length of v, namely, the square root of $v \cdot v$. If C is a finite set, then let $| C |$ be the number of elements in C. By $v \geq 0$ and $v > 0$ we mean $v^i \geq 0$ for $i = 1, \cdots, n$ and $v^i > 0$ for $i = 1, \cdots, n$.

By the convex hull of a set C in R^n we mean the set of all x in R^n which can be expressed as

$$x = \sum_{i \in N} \lambda_i x_i$$

where the x_i are in C and the λ_i's are nonnegative and satisfy

$$1 = \sum_{i \in N} \lambda_i.$$

Given a convex set C in R^n we denote its relative boundary by ∂C; hence, if C is n-dimensional, that is, if C has an interior in R^n, then ∂C and the boundary of C in R^n coincide.

Let u_i for $i \in N$ be the vertices of the n-simplex S in R^n; S is the convex hull of its vertices. In fact each $x \in S$ can be expressed uniquely as

$$x = \sum_{i \in N} \lambda_i(x) u_i \qquad 1 = \sum_{i \in N} \lambda_i(x)$$

where $\lambda_i(x) \geq 0$. The weight $\lambda_i(x)$ is called the ith barycentric coordinate of x in S. If $I \subseteq N$, then the convex hull of $\{u_i : i \in I\}$ is called the I-face of S or merely a face of S; if I contains exactly n elements, then the face is also called the I-facet of S.

The following four lemmas are used only in the examples of the next section, nevertheless, they are elementary and of general interest.

LEMMA 1: *Given $l : S \to R^n$ and a subset C of the n-simplex S, if there exists a vector v in R^n such that $v \cdot l(y) > 0$ for all $y \in C$, then C is not l-complete. If C is finite and not l-complete then there exists a vector v in R^n such that $v \cdot l(y) > 0$ for all $y \in C$.*

Proof: Suppose such a v exists and that C is l-complete; we have;

$$0 = \sum_{i \in N} \lambda_i l(x_i) \qquad 1 = \sum_{i \in N} \lambda_i$$

with $\lambda_i \geqq 0$. Taking an inner product with the origin and v we get

$$0 < \sum_{i \in N} \lambda_i v \cdot l(x_i) = 0$$

which is a contradiction and the "only if" half of the proof is established.

If C is not l-complete then the origin is not in the convex hull of $l(C)$. Since C is finite the convex hull of $l(C)$ is compact and contains an element v with minimum $\| v \|$. Using convexity it follows that $v \cdot l(y) > 0$ for $y \in C$. $\square$

LEMMA 2: *Let $l : S \to R^n$ be a proper labeling and let $\alpha : S \to (0, +\infty)$ be a positive function. Then $l' : S \to R^n$ defined by*

$$l'(x) = \alpha(x) l(x)$$

is also a proper labeling, and the complete sets and points of l and l' coincide.

Proof: Consider the system

$$0 = \sum_{i \in N} \lambda_i l_i \qquad 1 = \sum_{i \in N} \lambda_i.$$

Next let us scale l_i by $\alpha_i > 0$. We get

$$0 = \sum_{i \in N} \lambda_i' (\alpha_i l_i) \qquad 1 = \sum_{i \in N} \lambda_i'$$

where

$$\lambda_j' = (\lambda_j/\alpha_j) \left(\sum_{i \in N} (\lambda_i/\alpha_i) \right)^{-1}.$$

Hence we see that l-complete sets and l'-complete sets coincide. $\square$

Lemmas 2 and 3 are of a common nature. Given $H \subseteq R^{n+1}$ define the nonnegative cone of H to be the set of points of form

$$\sum_{i \in N} h_i x_i$$

where $h_i \in H$ and $0 \leq x_i \in R^1$ for $i \in N$, namely, it is the set of nonnegative linear combinations of elements of H. In Lemma 3 l^i, h^k, and b^k denote the ith and kth components of the corresponding vector.

LEMMA 3: *Given* $h:S \to R^{n+1}$ *and* $b \in R^{n+1}$ *suppose there is a vector* $v \in R^{n+1}$ *such that* $v \cdot h(x) > 0$ *for* $x \in S$. *Select* $j = 1, \cdots, n + 1$ *with* $v^j \neq 0$ *and define* $l:S \to R^n$ *by*

$$l^i(x) = \left(\frac{v \cdot b}{v \cdot h(x)} \right) h^k(x) - b^k, \quad \text{where } k = \begin{cases} i & i < j \\ & \text{for} \\ i + 1 & i \geq j \end{cases}.$$

Then a subset C *of* S *is* l-complete if and only if b is contained in the nonnegative cone of $h(C)$.*

Proof: If $v \cdot b = 0$ the result is trivial. Assume $v \cdot b > 0$ and $h_i \in h(C)$ for $i \in N$. Then $\sum_{i \in N} h_i x_i = b$ with $x_i \geq 0$ for $i \in N$ holds if and only if

$$\sum_{i \in N} \left(\left(\frac{v \cdot b}{v \cdot h_i} \right) h_i - b \right) \lambda_i = 0$$

and $\sum_{i \in N} \lambda_i = 1$ with $\lambda_i \geq 0$ for $i \in N$ holds where

$$\lambda_i = \left(\frac{v \cdot h_i}{v \cdot b} \right) x_i.$$

This establishes the "if" conclusion of the lemma.

On the other hand now suppose that

$$\sum_{i \in N} \left(\left(\frac{v \cdot b}{v \cdot h_i} \right) h_i^k - b^k \right) \lambda_i = 0 \qquad \text{for} \quad k \neq j$$

$\sum_{i \in N} \lambda_i = 1$, and $\lambda_i \geqq 0$ for $i \in N$. Multiply the kth term by v^k and sum over $k \neq j$ to get

$$\sum_{i \in N} \left(\left(\frac{v \cdot b}{v \cdot h_i} \right) (v \cdot h_i - v^j h_i^j) - (v \cdot b - v^j b^j) \right) \lambda_i = 0$$

or that

$$v^j \sum_{i \in N} \left(\left(\frac{v \cdot b}{v \cdot h_i} \right) h_i^j - b^j \right) \lambda_i = 0.$$

Dividing by v^j and using the "if" argument the lemma is established. $\square$

LEMMA 4: *Given* $l: S \to R^n$, *if* $x + l(x)$ *is interior to S for each x in the boundary of S, then l is a proper labeling.*

Proof: Suppose that the vertex set $\{u_i : i \in N\}$ of S is not l-complete. Then according to Lemma 1 there is a v such that $v \cdot l(u_i) > 0$ for all $i \in N$. Now choose $j \in N$ so that

$$v \cdot u_j = \max_{i \in N} v \cdot u_i = \max_{x \in S} v \cdot x.$$

Then we have

$$v \cdot (l(u_j) + u_j) > v \cdot u_j$$

which implies that $l(u_j) + u_j$ is not in S. This is a contradiction and it follows that the vertex set of S is complete.

On the other hand, now let v be normal to the hyperplane spanned by an I-facet of S and choose the sign of v so that $v \cdot u_j > v \cdot u_i$ for $j \notin I$ and $i \in I$. It follows that $v \cdot (x + l(x)) > v \cdot x$ or that $v \cdot l(x) > 0$ for each x in the I-facet; it follows that no facet of S is l-complete. $\square$

3. EXAMPLES

Here four examples of properly labeled simplexes are treated; the l-complete points yield, respectively, completely labeled simplexes in the sense of Sperner, fixed points in the sense of Kakutani, stationary points in the sense of optimization, and imputations in the core of a game.

Example 1, Sperner's Lemma: Our first example uses the notion of a triangulation of S; briefly a triangulation of S can be considered a collection T^n of n-simplexes in S such that

(i) The union of these n-simplexes, $\cup_{\sigma \in T^n}$, contains S,
(ii) any two of these n-simplexes σ and σ' in T^n meet in a face $\sigma \cap \sigma'$ (perhaps empty) of σ and of σ'.

The set T^0 of vertices of T^n is the set of all vertices of n-simplexes σ of T^n.

Let T^n be a triangulation of S. Let $f : T^0 \to N$ be an integer labeling of the set T^0 of vertices of the triangulation T^n. In addition, we require that if I is any subset of N, if $v \in T^0$ is a vertex of the triangulation, and if v is contained in the I-face of S, then $f(v)$ must lie in I. Sperner's Lemma concludes that there is an n-simplex σ in T^n such that f applied to the vertex set of σ yields N. See Figure 1. Our next step is to show that this result follows from Theorem 1.

Let $g(i)$ for $i \in N$ be vectors in R^n such that the origin is contained in the convex hull of $g(N)$ but such that the origin is not contained in the convex hull of any proper subset of $g(N)$, for example, let $(g(0), \cdots, g(n))$ be the columns of the $n \times (n + 1)$ matrix

$$\begin{pmatrix} 1 & -1 & 0 & & & 0 \\ 1 & 0 & -1 & & & \\ & & 0 & & & \\ & & \cdot & & & \\ & & \cdot & & & \\ & & \cdot & & & \\ & & \cdot & \cdots & & 0 \\ & & \cdot & & & \\ 1 & 0 & 0 & & & -1 \end{pmatrix}.$$

Define $l : S \to R^n$ as follows. For the vertices $v \in T^0$ let $l(v) =$

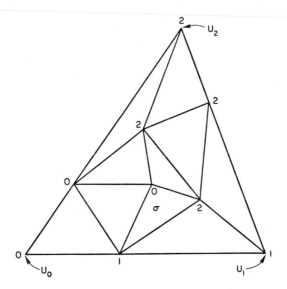

FIG. 1. A Simplex σ of a Triangulation of S with a Complete Set of Labels

$g(f(v))$. Then extend l barycentrically to all of S, that is, if x lies in the simplex $\sigma \in T$ with vertices $v_0, \cdots, v_n$, then let

$$l(x) = \sum_{i \in N} \lambda_i(x) l(v_i)$$

where $\lambda_i(x)$ is the ith barycentric coordinate of x in σ, that is

$$x = \sum_{i \in N} \lambda_i(x) v_i \qquad 1 = \sum_{i \in N} \lambda_i(x).$$

Upon applying l to the vertex set of S and to the $(N \sim \{i\})$-facet of S one gets $g(N)$ and a subset of the convex hull of $g(N \sim \{i\})$, respectively. Hence due to our choice of g, S is properly labeled. Let x be a complete point. Since l is continuous we have

$$0 = l(x) = \sum_{i \in N} \lambda_i(x) l(v_i)$$

or that x lies in an n-simplex whose vector labels include all of $g(N)$, that is, whose integer labels include all of N.

Example 2, Kakutani Fixed Point Theorem: (See [9, 11].) Let
$F: C \to 2^c$ be a point to set map from C to nonempty convex subsets of C where C is an n-dimensional compact convex set in R^n.
Further let us assume that the graph of f in $C \times C$ is closed, that is,
$x_k \to x_\infty$, $y_k \to y_\infty$, and $y_k \in f(x_k)$ for $k = 1, 2, \cdots$ imply that
$y_\infty \in f(x_\infty)$. Using Theorem 1 we proceed to show that f has a
fixed point, that is, there is a point x in C for which $x \in f(x)$. Let
$S \subset R^n$ be an n-simplex containing C in its interior and define
$l: S \to R^n$ by letting $l(x)$ be an element of $f(x) - x = \{y - x:$
$y \in f(x)\}$ if $x \in C$ and be the element $c - x$ if $x \notin C$, where c is a
fixed interior point of C. By Lemma 4, l is a proper labeling and we
proceed to show that complete points are fixed points.

Suppose that $C_k = \{x_{0k}, \cdots, x_{nk}\} \to x$ is a sequence of complete
sets tending to x where $x_{ik} \in S$. We have

$$0 = \sum_{i \in N} l(x_{ik})\lambda_{ik} \qquad 1 = \sum_{i \in N} \lambda_{ik}$$

and $\lambda_{ik} \geq 0$ for all k. Taking a subsequence if necessary, we can assume that $\lambda_{ik} \to \lambda_i$. Let I denote those indices i such that $x_{ik} \in C$
for infinitely many k and let $J = N \sim I$. We now consider the
limit of our equations above to get

$$0 = \sum_I \lambda_i(h_i - x) + t(c - x) \qquad 1 = s + t$$

where $s = \sum_I \lambda_i$. If $s = 0$, then x is not interior to C; but we have
$c - x = 0$ which is a contradiction. Hence $s > 0$ and using properties of f we can rewrite our equations to get

$$0 = (h - x) + t/s(c - x) \qquad 1 = s + t$$

where $h \in f(x)$. If $t > 0$ we see that x is in the boundary of C hence

$$h = x + t/s(x - c)$$

is not in C which is a contradiction. Therefore $t = 0$ and we have
$x = h \in f(x)$.

If $f: S \to 2^S$ and $f(\partial S)$ is contained in the interior of S, then one
can set $l(x)$ to be an arbitrary element of $f(x) - x$ for all x in S
to get a proper labeling.

Example 3, Stationary Points in Optimization: (See [9, 14].)
Let $f: C \to R^n$ be a function on a closed convex set C. A point $x \in C$
is called a stationary point if $x \cdot f(x)$ is greater than or equal to
$y \cdot f(x)$ for all $y \in C$.

The reader can verify that if x is an optimal solution to the program

$$\text{maximize}: g(x)$$

$$\text{subject to}: x \in C$$

where $g: C \to R^1$ is differentiable, then x is a stationary point of
$\nabla g: C \to R^n$, where ∇g is the gradient of g. Further if g is a concave
differentiable function and x is a stationary point of $\nabla g: C \to R^n$,
then x is optimal.

Now consider the task of computing a stationary point of $f: C \to$
R^n where f is continuous and C is compact convex, and n-dimensional. As in Example 2 let S be an n-simplex containing C in its
interior and let c be a point interior to C. Define $l: S \to R^n$ as follows: If $x \in C$ let $l(x) = f(x)$, if $x \in \partial S$ let $l(x) = c - x$, and if
$x \notin C \cup \partial S$ let $0 \neq l(x) \in R^n$ be a vector v such that $y \cdot v \geqq x \cdot v$
for all $y \in C$. From Lemma 4 $l: S \to R^n$ is a proper labeling and we
proceed to show that l-complete points are stationary points.

We can assume via Lemma 2 that $\| l(x) \| = 1$ for $x \notin C \cup \partial S$.
Let $\{x_{0k}, \cdots, x_{nk}\} \to x$ be a sequence of l-complete sets tending to
x. We have

$$0 = \sum_{i \in N} l(x_{ik}) \lambda_{ik} \qquad 1 = \sum_{i \in N} \lambda_{ik}$$

and $\lambda_{ik} \geqq 0$. Taking subsequences if necessary we can assume that
$\lambda_{ik} \to \lambda_i$ as $k \to +\infty$ for each i. Let I be those indices i such that
$x_{ik} \in C$ for infinitely many k and $J = N \sim I$.

First suppose that I is empty; we shall obtain a contradiction
by showing that $v \cdot l(x_{ik})$ is positive where $v = c - x$ for $i \in N$
with sufficiently large k and by using Lemma 1. Certainly $(c - x) \cdot$
$(c - x_{ik})$ is positive for large k since x is not interior to C. If
$y \cdot v_{ik} \geqq x_{ik} \cdot v_{ik}$ for $y \in C$ where $\| v_{ik} \| = 1$, then by letting $d > 0$ be
the distance from c to the boundary of C we get

$$c \cdot v_{ik} \geqq x_{ik} \cdot v_{ik} + d$$

or that
$$(c - x_{ik}) \cdot v_{ik} \geqq d$$
or that
$$(c - x) \cdot v_{ik} \geqq (\tfrac{1}{2})d$$

for sufficiently large k. Hence we see that I is not empty and hence $x \in C$.

Taking the limit in our equations above and using the properties of f we get
$$0 = sf(x) + tv \qquad 1 = s + t$$
where $s = \sum_{i \in I} \lambda_i$, $\| v \| = 1$, and $y \cdot v \geqq x \cdot v$ for $y \in C$. Clearly $s > 0$, hence
$$f(x) = -(t/s)v$$
demonstrating that x is a stationary point. $\square$

Example 4, Scarf's Theorem on the Core of a Game: We present here a theorem of Scarf [3] as generalized by Billera [15] which states that a balanced cooperative game without side payments has a nonempty core. Our method of proof regarding the game follows that of Shapley [16].

Regard R^N as R^{n+1} with axes $e_0 = (1, 0, \cdots, 0), \cdots, e_n = (0, \cdots, 0, 1)$; if $x \in R^N$ then $x = (x^0, x^1, \cdots, x^n)$. For $I \subseteq N$ let R^I and R_+^I be the points x and $x \geqq 0$ in R^N such that $x^i = 0$ for $i \in N \sim I$. Let N^* be the set of $I \subseteq N$ for which $I \neq \phi$ and $I \neq N$.

Consider an $(n + 1)$-person game with players $i \in N$. For each (nonempty) coalition of players $I \subseteq N$ let $V_I \subseteq R^N$ be a set representing the capability of the coalition acting alone (regardless of the behavior of players $N \sim I$). In particular if $x = (x^0, \cdots, x^n) \in V_I$, then the coalition can secure the amount x^i for each member $i \in I$.

We assume that each player acting alone can guarantee himself the amount zero, that is, we assume $V_i = \{x \in R^N : x^i \leqq 0\}$. In addition for each coalition I we assume that $V_I \neq \phi$ and

(1) $x \in V_I$ and $y \leqq x$ imply $y \in V_I$,
(2) V_I is closed,
(3) $V_I + R^{N \sim I} = V_I$,
(4) $V_I \cap R_+^I$ is bounded.

Although we shall not, these conditions can be given a plausible justification in game parlance.

The core of the game is defined to be the set

$$V_N \sim (\text{interior } (V \cup V_N))$$

where $V = \cup_{I \in N^*} V_I$. It is the set of payoffs x achievable by the grand coalition N which cannot be blocked, that is, $x \in V_N$ is in the core if and only if there does not exist an $I \subset N$ and $y \in V_I$ with $y^i > x^i$ for $i \in I$, or equivalently in view of (3), with $y > x$.

Clearly the core might be empty and we continue by introducing the concept of a balanced game. Let $0 < b \in R^N$ be a positive vector and let $g:N^* \to R_+^N$ be a function such that $0 \leqq g(I) \neq 0$ and such that $(g(I))^i > 0$ implies $i \in I$ for all $I \in N^*$.

A collection of coalitions $\beta \subseteq N^*$ is said to be balanced if b is contained in the nonnegative cone of $g(\beta)$, see Section 2. If each balanced collection $\beta \subseteq N^*$ has the property that

$$V_N \supseteq \bigcap_{I \in \beta} V_I$$

then we say the game is balanced. Using Theorem 1 we proceed to show that a balanced game has a nonempty core.

Define M to be a number exceeding $n\bar{M}$ where $\bar{M} = \sup\{x^i : x \in V_I, i \in I, I \in N^*\}$. Due to conditions (3) and (4) we see that finite $M > n\bar{M} \geqq 0$ exist. Let $u_0 = (0, M, \cdots, M)$, $u_1 = (M, 0, M, \cdots, M)$, $\cdots u_n = (M, \cdots, M, 0)$ be the vertices of S.

Define $\mathscr{s}:S \to N^*$ by setting $\mathscr{s}(x)$ to be the set of coalitions $I \in N^*$ such that

$$\max\{\lambda : x + \lambda e \in V_I\} = \max\{\lambda : x + \lambda e \in V\}.$$

These maximums are always attained due to the facts regarding the V_I's. Clearly $\mathscr{s}(x)$ is nonempty for $x \in S$.

Our interest in $\mathscr{s}$ lies in the fact that if $\mathscr{s}(y)$ contains a balanced collection, then $y + \lambda e$ is in the core where λ is maximum subject to $y + \lambda e \in V \cup V_N$. To see this observe that if $y + \lambda e < x$ for $x \in V_J$, then $y + \lambda e$ is interior to $V \cup V_N$ which is a contradiction to the definition of λ. Further, $y + \lambda e$ is in V_N since the game is balanced. To show that such a y exists we must establish two more properties of $\mathscr{s}$.

First, if x is in the I-face of S, then $J \in g(x)$ implies that $J \subseteq I$. To see this observe that $x^k = M$ for $k \notin I$ and that for some $i \in I$ we have $x^i \leq (1 - 1/n)M$, hence setting $\mu = -(1 - 1/n)M$ we get $(x + \mu e)^i \leq 0$ and $(x + \mu e)^k = M - (1 - 1/n)M = 1/nM > \bar{M}$. It follows that $(x + \lambda e) \notin V_J$ for $\lambda \geq \mu$ and $(x + \lambda e) \in V_i$ for $\lambda = \mu$.

Second, g has a closed graph, that is, $x_k \to x_\infty$ and $I \subset g(x_k)$, as $k \to \infty$ imply $I \subseteq g(x_\infty)$. From (1) we can conclude for any x and y in S that

$$| \max\{\lambda : y + \lambda e \in V_I\} - \max\{\lambda : x + \lambda e \in V_I\} | \leq \| x - y \|$$

or that

$$x \to \max\{\lambda : x + \lambda e \in V_I\}$$

is continuous. Hence if

$$\max\{\lambda : x_k + \lambda e \in V_I\} = \max\{\lambda : x_k + \lambda e \in V_I, I \in N^*\}$$

for $k = 1, \cdots$, then the equation also holds for $k = \infty$.

Define $h : S \to R^n$ by setting $h(x) = g(I)$ for some $I \subseteq g(x)$. Using Lemma 4 with $v = e$ we can regard a subset of C of S as complete if and only if b is in the nonnegative cone of $h(C)$. Using the property we have established regarding the behavior of g on faces of S, we see that h applied to the vertex set of S yields the vectors $g(i)$ for $i \in N$; hence the vertex set is complete. Further, if x is in the $(N \sim \{i\})$-facet of S we see that $(h(x))^i = 0$ and hence that the $(N \sim \{i\})$-facet is not complete. Therefore, we have a proper labeling and there is a complete point y. Using the fact that g has a closed graph we see that $g(y)$ contains a balanced collection.

4. GRADUATED PSEUDOMANIFOLDS ON $S \times [0, +\infty)$

To describe and prove convergence of our algorithm it is convenient if not necessary to introduce another dimension, that is, to consider the cylinder $S \times [0, +\infty)$ in R^{n+1}. The algorithm generates an infinite sequence $\sigma_0, \sigma_1, \cdots$ of distinct sets of size $n + 1$ in the cylinder. The sets σ_i and σ_{i+1} differ by one element and the sequence σ_i tends to ∞ in the cylinder. A sequence of l-complete sets C_k, $k = 0, 1, 2, \cdots$, is obtained by projecting the sequence σ_k, $k = 0, 1, 2, \cdots$, to S. The pairs C_k and C_{k+1} will differ by at most one element and it is desired that the diameter of the C_k tend to zero. These considerations lead to the notion of a graduated pseudomanifold on the cylinder.

Given a set σ in the cylinder $S \times [0, +\infty)$, we define $\pi_1 : S \times [0, +\infty) \to S$ to be the projection to S, namely,

$$\pi_1(\sigma) = \{x \in S : (x, t) \in \sigma\}.$$

We define $\pi_2 : S \times [0, +\infty) \to [0, +\infty)$ to be infimum of the projection to $[0, +\infty)$, namely,

$$\pi_2(\sigma) = \inf\{t \in [0, +\infty) : (x, t) \in \sigma\}.$$

By a facet of the cylinder we mean $S \times \{0\}$ or $T \times [0, +\infty)$, where T is a facet of S.

Let K^{n+1} be a collection of subsets τ of the cylinder $S \times [0, +\infty)$ of size $n + 2$, that is, $\tau \in K^{n+1}$ implies $\tau \subseteq S \times [0, +\infty)$ and $|\tau| = n + 2$. For $i = 0$ and $i = n$ let $K^i = \{\sigma : \sigma \subset \tau \in K^{n+1}, |\sigma| = i + 1\}$ be the collection of all subsets of elements of K^{n+1} of size $i + 1$. We call elements of K^n and K^{n+1} (abstract) n and $(n + 1)$-simplexes, respectively, and elements of K^0 (abstract) vertices. If $\sigma \subset \tau$, where σ is an n-simplex of K^n and τ is an $(n + 1)$-simplex of K^{n+1}, we call σ a (abstract) facet of τ.

DEFINITION: K^{n+1} is defined to be a graduated pseudomanifold on the cylinder $S \times [0, +\infty)$ if and only if the following five conditions hold:

(i) The set $\sigma_0 = \{(u_i, 0) : i \in N\}$ of vertices of $S \times \{0\}$ is an n-simplex of K^n. Further σ_0 is the unique n-simplex of K^n contained in $S \times \{0\}$.

(ii) Each n-simplex σ of K^n is a facet of exactly one or two $(n + 1)$-simplexes τ of K^{n+1}.

(iii) An n-simplex σ of K^n is a facet of one $(n + 1)$-simplex τ of K^{n+1} if and only if σ is contained in a facet of the cylinder.

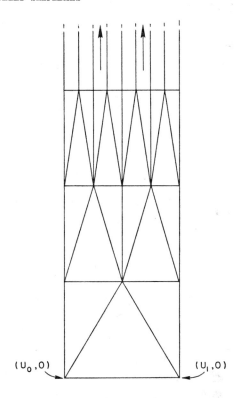

Fig. 2. A Graduated Pseudomanifold on $S \times [0, +\infty)$ with $S = [u_0, u_1]$

(iv) For each $t \in [0, +\infty)$ the set of n-simplexes $\sigma \in K^n$ with $\pi_2(\sigma) \leqq t$ is finite.

(v) If $\pi_2(\sigma_i) \rightarrow +\infty$ for an infinite sequence of n-simplexes σ_i in K^n, then the diameter of $\pi_1(\sigma_1)$ tends to zero. □

Figure 2 illustrates a graduated pseudomanifold K^2 on the cylinder $S \times [0, +\infty)$ where S is a 1-simplex. The (abstract) 2-simplexes of K^2 are those 2-simplexes which are evident in the figure. Item (i) is related to initiating the algorithm; in fact σ_0 is the first member of the generated sequence $\sigma_0, \sigma_1, \cdots$ of elements in K^n. Items (ii) and (iii) bear on the result that the sequence $\sigma_0, \sigma_1, \cdots$ is infinite and distinct. Item (iv) forces the sequence

σ_0, σ_1, $\cdots$ to rise in the cylinder. Item (v) forces the diameter of σ_i projected to S to tend to zero.

A triangulation of the cylinder $S \times [0, +\infty)$, in the classical sense (e.g., as in Figure 2), for which (i) and (v) obtain yields a graduated pseudomanifold on the cylinder. The convergence rate of our algorithm is highly dependent on the choice of graduated pseudomanifold, hence, we did not define graduated pseudomanifolds in terms of triangulation due to the otherwise unnecessary reduction in options and flexibility.

For each n there is a decided variety of graduated pseudomanifolds on $S \times [0, +\infty)$; we shall not prove this. To actually implement the algorithm one needs a formula for generating the graduated pseudomanifold; see K_1 and K_2 of [13] where a pair of such formulas is given.

5. LABELING K^0 AND VERY COMPLETE SETS

Let $l: S \to R^n$ be a proper labeling, let K^{n+1} be a graduated pseudomanifold on the cylinder $S \times [0, +\infty)$, and let K^0 and K^n be the vertices and facets of K^{n+1}. We define $L: K^0 \to R^n$ by setting $L(x, t) = l(x)$ for each vertex $(x, t) \in K^0$. Our next step is to introduce a notion which is slightly stronger than an l-complete set; this device, an L-very complete set, is necessary to prevent the algorithm from cycling.

For $\epsilon \in R^1$ let $[\epsilon]$ denote the vector $(\epsilon, \epsilon^2, \cdots, \epsilon^n)$ in R^n of powers of ϵ.

DEFINITION: An n-simplex $\sigma \in K^n$ is defined to be L-very complete if and only if $[\epsilon]$ is contained in the convex hull of $L(\sigma)$ for all sufficiently small $\epsilon > 0$. $\square$

Clearly if $\sigma \in K^n$ is L-very complete then its projection to S, $\pi_1(\sigma)$, is l-complete.

LEMMA 5: *The n-simplex σ_0 in K^n, that is the vertex set of $S \times \{0\}$, is L-very complete.*

Proof: The vertex set $\{u_1 : i \in N\}$ of S is l-complete by (a) in the definition of a proper labeling. Hence the system

$$\sum_{i \in N} \binom{1}{l_i} \lambda_i = \binom{1}{0}$$

has a nonnegative solution in λ_i, where $l_i = L(u_i, 0) = l(u_i)$. Since no proper face of S is l-complete it follows that each λ_i must be positive in the solution. As a consequence we see that the $(n + 1) \times (n + 1)$ matrix

$$\begin{pmatrix} 1 & & 1 \\ l_0, & \cdots, & l_n \end{pmatrix}$$

has an inverse and that the vector

$$\begin{pmatrix} 1 & & 1 \\ l_0, & \cdots, & l_n \end{pmatrix}^{-1} \binom{1}{0}$$

is positive. Further it follows that for sufficiently small $\epsilon > 0$

$$\begin{pmatrix} 1 & & 1 \\ l_0, & \cdots, & l_n \end{pmatrix}^{-1} \binom{1}{[\epsilon]}$$

is positive. Therefore σ_0 is L-very complete. $\square$

LEMMA 6: *If the n-simplex σ of K^n is L-very complete and σ is a facet of τ of K^{n+1}, then τ has exactly one other L-very complete facet.*

Proof: First one observes that for any $\epsilon_0 > 0$ the set $\{(1, [\epsilon]) : 0 < \epsilon \leqq \epsilon_0\}$ contains a basis of R^{n+1}. This can be proved by induction on n by showing that the determinant of the $(n + 1) \times (n + 1)$ matrix with columns $(1, [\epsilon_i])$ for $i \in N$ for some $0 < \epsilon_n < \epsilon_{n-1} < \cdots < \epsilon_0$ is not zero; one computes determinants by cofactors by expansion in the row of 1's.

Second, let $v_0, \cdots, v_{n+1}$ be the vertices of τ, let $v_0, \cdots, v_n$ be the vertices of σ, and let $l_i = L(v_1)$ for $i = 0, \cdots, n + 1$. For all sufficiently small $\epsilon > 0$ we know that the system of equations

$$\sum_{i \in N} \binom{1}{l_i} \lambda_i + \binom{1}{l_{n+1}} \theta = \binom{1}{[\epsilon]}$$

has a nonnegative solution in λ_i with $\theta = 0$ (since σ is L-very complete). We shall complete the proof by showing that there is a unique $k \in N$ such that this system has a nonnegative solution in λ_i and θ with $\lambda_k = 0$ for all sufficiently small $\epsilon > 0$. Since $(1, [\epsilon])$ spans R^{n+1} for small ϵ we know that the matrix

$$\left(\binom{1}{l_0}, \ldots, \binom{1}{l_n} \right)$$

has an inverse. We premultiply the system of equations above by this inverse to get the system of equations

$$\lambda_i = p_i(\epsilon) - a_i \theta \qquad i \in N$$

with the same solution set in the λ_i and θ.

Since the solution set of this system is bounded for any ϵ we see that the set $I = \{i \in N : a_i > 0\}$ is nonempty. Since rows of an invertible matrix are linearly independent it follows that the polynomials $p_i(\epsilon)$ for $i \in N$ are nontrivial. Since each of these polynomials has only finitely many zeros we see that there is an $\epsilon_0 > 0$ such that each polynomial $p_i(\epsilon)$ is positive for $0 < \epsilon \leq \epsilon_0$. Repeating this line of reasoning it follows that the polynomials $p_i(\epsilon)/a_i$ for $i \in I$ are distinct, and hence for any two members i and j of I there is an $0 < \epsilon_{ij} \leq \epsilon_0$ such that $p_i(\epsilon)/a_i \neq p_j(\epsilon)/a_j$ for $0 \leq \epsilon \leq \epsilon_{ij}$. Letting $\epsilon_1 = \min \epsilon_{ij}$ we see that there is a unique $k \in I$ such that

$$p_k(\epsilon)/a_k < p_i(\epsilon)/a_i$$

for all $i \in I \sim \{k\}$ and $0 < \epsilon \leq \epsilon_1$.

To get another L-very complete facet of τ, we need a solution to the system of equations with θ positive, some fixed λ_j zero, and the remaining λ_i nonnegative for all small $\epsilon > 0$. It follows that θ must have the value

$$\theta_0 = \min_{i \in I} p_i(\epsilon)/a_i$$

for any $0 < \epsilon \leqq \epsilon_0$. Note that for $0 \leqq \theta < \theta_0$ we have $\lambda_i > 0$ for all $i \in N$ and for $\theta > \theta_0$ we have $\lambda_j < 0$ for some $j \in N$. Hence setting $\theta = p_k(\epsilon)/a_k > 0$ and $\lambda_i = p_i(\epsilon) - a_i\theta$ for $i \in N$ for $0 < \epsilon \leqq \epsilon_1$ we get the unique other nonnegative solution with a zero component; $\sigma' = \tau \sim \{v_k\}$ is the one other L-very complete facet of τ. □

DEFINITION: Two distinct n-simplexes of K^n are defined to be adjacent if they are both facets of an $(n + 1)$-simplex in K^{n+1}. □

LEMMA 7: *The vertex set σ_0 of $S \times \{0\}$ is adjacent to exactly one L-very complete simplex of K^n. Every other L-very complete simplex in K^n is adjacent to exactly two L-very complete simplexes of K^n.*

Proof: By Lemma 5 σ_0 is L-very complete and by (i) and (iii) in the definition of a graduated pseudomanifold σ_0 is a facet of exactly one simplex τ of K^{n+1}. By Lemma 6 τ has exactly two $\sigma_1 \neq \sigma_0$ L-very complete facets. Hence by definition of adjacency σ_1 is the unique L-very complete simplex of K^n that is adjacent to σ_0. If $\sigma \neq \sigma_0$ is L-very complete, then σ is not in a facet of the cylinder because by (i) in the definition of a graduated pseudomanifold σ is not in the facet $S \times \{0\}$ and by (b) in the definition of a proper labeling σ is not in a facet $T \times [0, +\infty)$. In view of condition (ii) σ is contained in two, $\tau_1 \neq \tau_2$, $(n + 1)$-simplexes of K^{n+1}. By Lemma 6 each τ_i contains exactly one other L-very complete simplex. It follows that σ is adjacent to exactly two L-very complete simplexes of K^n. □

In actual computation one never manipulates ϵ as used in this section; rather the tasks of obtaining L-very complete sets are accomplished by making lexicographic comparisons of vectors of ratios just as in the simplex method.

6. THE ALGORITHM

We are now prepared to specify our algorithm. Let $l: S \to R^n$ be a proper labeling and let K^{n+1} be a graduated pseudomanifold on the cylinder with facets K^n and vertices K^0. The algorithm generates an infinite sequence σ_0, σ_1, $\cdots$ of distinct adjacent L-very complete n-simplexes in K^n. The projection of this sequence to S, $C_i = \pi_1(\sigma_i)$, yields a sequence C_0, C_1, $\cdots$ of l-complete sets in S whose diameters tend to zero. Since S is compact the sequence C_0, C_1, $\cdots$ will have limit points, further these limit points form a connected set.

The Algorithm:

Step 0: Let $\sigma_{-1} = \sigma_0$ be the vertex set of $S \times \{0\}$.

Step $k \geqq 1$: Assuming σ_0, σ_1, $\cdots$, σ_{k-1} have been generated, let $\sigma_k \neq \sigma_{k-2}$ be the unique L-very complete simplex of K^n adjacent to σ_{k-1}. $\square$

THEOREM 2: *The algorithm generates a unique infinite sequence of distinct adjacent L-very complete n-simplexes in K^n.*

Proof: By Lemma 5 σ_0 is L-very complete. By Lemma 7 σ_0 is adjacent to exactly one L-very complete simplex; this must be σ_1. Assume that the sequence σ_0, $\cdots$, σ_{k-1} for some $k \geqq 2$ has been generated by the algorithm and is adjacent, unique, and distinct. Since $\sigma_{k-1} \neq \sigma_0$, by Lemma 7 σ_{k-1} is adjacent to exactly two L-very complete n-simplexes, σ and σ_{k-2}, hence, σ_k must be σ. Suppose $\sigma_0 = \sigma_k$, then $k > 2$. However since σ_0, $\cdots$, σ_{k-1} is distinct and adjacent and since σ_0 is adjacent only to σ_1, we have $k = 2$ which is a contradiction. If $\sigma_k = \sigma_i$ for some $0 < i < k - 2$, then σ_{k-1} must be σ_{i-1} or σ_{i+1} since σ_i is adjacent only to two L-very complete simplexes. This contradicts the distinctness of the sequence σ_0, $\cdots$, σ_{k-1}. It follows that the sequence σ_0, $\cdots$, σ_k is adjacent and distinct. The result follows by induction. $\square$

Figure 3 illustrates this process where $n = 1$; an example in higher dimension would provide more insight but such an example does not lend itself to a drawing.

Given a sequence C_0, C_1, $\cdots$ of subsets of S, then we define the set of limit points of the sequence to be all those points x in S such that x is a limit point of a (infinite) subsequence C_i, $i \in K$. We say that a subset T of S is ϵ-connected if for any two points s_1 and

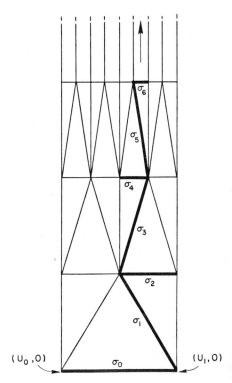

FIG. 3. The Unique Infinite Sequence $\sigma_0, \sigma_1, \cdots$ of Distinct Adjacent L-Very Complete 1-Simplexes with $S = [u_0, u_1] = [0, 1]$ and $l(x) = 1 - 2x^2$

s_2 of T there exist points $s_1 = t_0, t_1, \cdots, t_m = s_2$ in T for some m such that $\| t_i - t_{i+1} \| \leqq \epsilon$ for $i = 0, \cdots, m - 1$. It is easily verified that a compact T is connected if and only if T is ϵ-connected for every $\epsilon > 0$.

THEOREM 3: *Let* $l : S \to R^n$ *be a proper labeling,* K^{n+1} *be a graduated pseudomanifold on the cylinder* $S \times [0, +\infty)$, $\sigma_0, \sigma_1, \cdots$ *be the sequence generated by the algorithm, and* $C_i = \pi_1(\sigma_i)$ *be the projection of* σ_i *to* S *for* $i = 0, 1, \cdots$. *The set of limit points of the sequence* $C_0, C_1, \cdots$ *is a nonempty compact connected set of l-complete points of* S.

Proof: By Theorem 2 the sets σ_i are distinct. By (iv) in the definition of a graduated pseudomanifold we see that $\pi_2(\sigma_i) \to +\infty$ as $i \to +\infty$. By (v) the diameter of C_i tends to zero. From the compactness of S it follows that the set of limit points is nonempty and compact. Each such point is an l-complete point by definition. Since the diameters of the C_i tend to zero and since C_i and C_{i+1} for $i = 0, 1, \cdots$ differ by at most one point it is clear that the set of limit points is ϵ-connected for each $\epsilon > 0$ and hence that the set of limit points is connected. $\quad\square$

Theorem 1 follows from Theorem 3.

REFERENCES

1. Lemke, C. E., and J. T. Howson, Jr., "Equilibrium points of bimatrix games," *Journal of the Society of Industrial Applied thematics* 12, **2** (1964), 413–423.
2. Lemke, C. E., "Bimatrix equilibrium points and mathematical programming," *Management Science* 11, 7 (1965), 681–689.
3. Scarf, H., "The core of an N person game," *Econometrica* 35, **1** (1967), 50–69.
4. ———, "The approximation of fixed points of a continuous mapping," *SIAM J. Appl. Math.* 15, **5** (1967), 1328–1343.
5. ———, *Ten Economic Studies in the Tradition of Irving Fisher*, New York: Wiley, 1967.
6. Cohen, D. I. A., "On the Sperner lemma," *J. Combinatorial Theory*, **2** (1967), 585–587.
7. Fan, K., "Simplicial Maps from an orientable n-pseudomanifold into S^m with the octahedral triangulation," *J. Combinatorial Theory* 2, **4** (1967), 588–602.
8. Kuhn, H. W., "Simplicial approximation of fixed points," *Proc. Nat. Acad. Sci.*, **61** (1968), 1238–1242.
9. Hansen, T., and H. Scarf, "On the applications of a recent combinatorial algorithm," (Cowles Foundation Discussion Paper No. 272, April 1969).
10. Eaves, B. C., "An odd theorem," *Proc. Amer. Math. Soc.* 26, **3** (1970), 509–513.
11. ———, "Computing Kakutani fixed points," *SIAM J. Appl. Math.* 21, **2** (1971), 236–244.
12. ———, "On the basic theorem of complementarity," *Mathematical Programming* 1, **1** (1971), 68–75.
13. ———, "Homotopies for computation of fixed points," *Mathematical Programming* 3, **1** (1972), 1–22.

14. ———, "Nonlinear programming via kakutani fixed points," Working No. 294, Center for Research in Management Science, University of California, Berkeley, California (1970).

15. Billera, L. J., "Some theorems on the core of an n-person game without side payments," *SIAM J. Appl. Math.*, **18** (1970), 567–579.

16. Shapley, L. S., "On balanced games without side payments," *Mathematical Programming*, edited by T. C. Hu and S. M. Robinson, New York: Academic Press, 1973.

17. Casper, "Assuming the haunted house has only finitely many rooms, there is a room without a ghost. The argument supporting this conclusion is similar to that used in the proof of Theorem 2," private communication. Also see Fan [7].

BOTTLENECK EXTREMA

Jack Edmonds
and
D. R. Fulkerson

0. ABSTRACT

Let E be a finite set. Call a family of mutually noncomparable subsets of E a clutter on E. It is shown that for any clutter $\mathfrak{R}$ on E, there exists a unique clutter $\mathfrak{S}$ on E such that, for any function f from E to real numbers,

$$\min_{R \in \mathfrak{R}} \max_{x \in R} f(x) = \max_{S \in \mathfrak{S}} \min_{x \in S} f(x).$$

Specifically, $\mathfrak{S}$ consists of the minimal subsets of E that have non-empty intersection with every member of $\mathfrak{R}$. The pair $(\mathfrak{R}, \mathfrak{S})$ is called a blocking system on E. An algorithm is described and several examples of blockings systems are discussed.

1. INTRODUCTION

Gross [7] has described an algorithm and a duality theorem for the bottleneck assignment problem: Given a square array of numbers, find a circling of entries with exactly one circle in each row and one circle in each column so as to maximize the value of the smallest circled entry. (For an interpretation, think of rows of the array as corresponding to men, columns to jobs, on a serial production line, with the entry in row i and column j being the rate at which man i can process items if he is assigned to job j.) An earlier, less efficient algorithm for this problem was given by Fulkerson, Glicksberg, and Gross [5]. The duality theorem proved by Gross is: Let $I = \{1, 2, \cdots, n\}$; let Π be the set of permutations of I; let $|C|$ denote cardinality of C, and let a_{ij} (for $i, j \in I$) be real numbers. Then

$$\max_{\pi \in \Pi} \min_{i \in I} a_{i,\pi(i)} = \min_{\substack{A, B \subset I \\ |A|+|B|=n+1}} \max_{\substack{i \in A \\ j \in B}} a_{ij}.$$

Similarly, the following bottleneck path problem has been considered by Pollack [12], Hu [9], and Fulkerson [4]. Let G be a network (graph) whose arcs (edges) have numerical "weights." Let a and b be two nodes (vertices) in G. Find in G a path P from a to b such that the minimum single arc-weight in P is a maximum. (For an interpretation, think of G as a flow-network with source a, sink b, where the weight of an arc is its flow-capacity.) The duality theorem noted in [4] for bottleneck paths is: The maximum of the minimum weight of an arc in a path from a to b is equal to the minimum of the maximum weight of an arc in a cut separating b from a. Here a *cut* separating b from a is a minimal set of arcs such that deleting them from G leaves a network which contains no path from a to b; "minimal" means that no proper subset has the same property. (If arcs in G are directed, "path" is interpreted to mean "uniformly directed path.")

The well-known traveling salesman problem is to find, in a given graph G whose arcs (possibly directed) have numerical weights, a minimum weight closed path that contains each node of G just once. A closed path that contains each node of G once is called a Hamilton *tour*. The bottleneck traveling salesman problem is to find a Hamilton tour such that the largest arc-weight in the tour is minimum. Gilmore and Gomory [6] have solved a special case of the traveling salesman problem and also a special case of the bottleneck traveling salesman problem.

The reader should now be able to pose bottleneck problems galore. For the moment, we give two more examples. In an undirected graph G whose arcs have weights, find a spanning tree T such that the maximum weight of an arc not in T is minimum. In an undirected graph G whose nodes have weights, find a set C of nodes such that C meets all of the arcs, and such that the maximum node-weight in C is minimum.

2. THE BOTTLENECK THEOREM AND THRESHOLD METHOD

Let E be a finite set. A *family* $\mathfrak{F}$ *on* E is a family of subsets of E. E is called the *domain* of $\mathfrak{F}$ (regardless of whether the union of members of $\mathfrak{F}$ is E). We define a *clutter* $\mathfrak{R}$ *on* E to be a family $\mathfrak{R}$ on E such that no member of $\mathfrak{R}$ is contained in another member of $\mathfrak{R}$.

The interest cited in bottleneck problems prompts the following theorem.

THEOREM: *For any clutter* $\mathfrak{R}$ *on a finite set* E, *there exists a unique clutter* $\mathfrak{S} = b(\mathfrak{R})$ *on* E *such that, for any function* f *from* E *to real numbers,*

$$(1) \qquad \min_{R \in \mathfrak{R}} \max_{x \in R} f(x) = \max_{S \in \mathfrak{S}} \min_{x \in S} f(x).$$

Specifically, $\mathfrak{S} = b(\mathfrak{R})$ *is the clutter consisting of the minimal subsets of* E *that have non-empty intersection with every member of* $\mathfrak{R}$.

COROLLARY: $b(b(\mathfrak{R})) = \mathfrak{R}$.

We call $\mathfrak{S}$ the *blocking clutter*, or simply the *blocker*, of $\mathfrak{R}$. By a *blocking system* on E we shall mean any two families $\mathfrak{R}$ and $\mathfrak{S}$ on E that satisfy (1) for every f, regardless of whether $\mathfrak{R}$ and $\mathfrak{S}$ are clutters.

If $\mathfrak{F}$ is any family on E, in place of $\mathfrak{R}$ and $\mathfrak{S}$, respectively, in (1), denote the left side of (1) as $u(\mathfrak{F}, f)$ and the right side of (1) as $w(\mathfrak{F}, f)$. The bottleneck problems, determine $u(\mathfrak{F}, f)$ and determine $w(\mathfrak{F}, f)$, where $\mathfrak{F}$ is any family on E, reduce to the case in which $\mathfrak{F}$ is a clutter on E, since clearly:

(2) *Where f is any real-valued function on E, $u(\mathfrak{F}, f) = u(\mathfrak{R}, f)$ and $w(\mathfrak{F}, f) = w(\mathfrak{R}, f)$ for any families $\mathfrak{F}$ and $\mathfrak{R}$ on E such that every member of $\mathfrak{F}$ has some member of $\mathfrak{R}$ as a subset and such that $\mathfrak{R} \subset \mathfrak{F}$.*

In particular, these equations hold if $\mathfrak{F}$ is arbitrary and $\mathfrak{R}$ consists of those members of $\mathfrak{F}$ that contain no other member of $\mathfrak{F}$.

Central to our subject is the following property for a pair $(\mathfrak{R}, \mathfrak{S})$ of families on E:

(3) *For any partition of E into two sets E_0 and E_1 ($E_0 \cap E_1 = \phi$ and $E_0 \cup E_1 = E$), either a member of $\mathfrak{R}$ is contained in E_0 or a member of $\mathfrak{S}$ is contained in E_1, but not both.*

The bottleneck theorem, above, follows immediately from Lemmas A, B, and C.

LEMMA A: *Any blocking system satisfies property* (3).

LEMMA B: *For any clutter $\mathfrak{R}$ on a set E, the $\mathfrak{S} = b(\mathfrak{R})$ specified in the theorem is the one and only clutter on E such that* (3) *holds.*

LEMMA C: *Any pair $(\mathfrak{R}, \mathfrak{S})$ of families on E satisfying* (3) *is a blocking system.*

The proof of Lemma C will be an algorithm, based on (3), for computing $u(\mathfrak{R}, f)$ and $w(\mathfrak{S}, f)$, thereby showing them to be equal. This algorithm, which we call the threshold method, requires only a small number of "steps," where each step consists mainly of deciding, for a given bipartition (E_0, E_1) of E, which of the two alternatives in (3) holds. Thus the threshold method is a good algorithm provided that there is a good algorithm for the latter.

Proof of Lemma A: That a blocking system satisfies (3) follows from equation (1), where $f(x) = 0$ for $x \in E_0$ and $f(x) = 1$ for $x \in E_1$. If the resulting value of $u(\mathfrak{R}, f) = w(\mathfrak{S}, f)$ is 0, then some member of $\mathfrak{R}$ is contained in E_0 and no member of $\mathfrak{S}$ is contained in E_1. If the resulting value of $u(\mathfrak{R}, f) = w(\mathfrak{S}, f)$ is 1, then no member of $\mathfrak{R}$ is contained in E_0 and some member of $\mathfrak{S}$ is contained in E_1.

Proof of Lemma B: It is convenient to consider another operator, $d(\mathfrak{R})$, defined for every clutter $\mathfrak{R}$ on E: $d(\mathfrak{R})$ consists of the complements in E of the members of $\mathfrak{R}$. Clearly $d(\mathfrak{R})$ is a clutter on E, and $d(d(\mathfrak{R})) = \mathfrak{R}$. Property (3) seems more transparent in terms of $\mathfrak{R}$ and the family $p(\mathfrak{R}) = d(b(\mathfrak{R}))$, and so it is useful to view $b(\mathfrak{R})$ as $d(p(\mathfrak{R}))$.

For any clutter $\mathfrak{R}$ on E, define $p(\mathfrak{R})$ to consist of the maximal subsets of E that contain no member of $\mathfrak{R}$. Clearly $p(\mathfrak{R})$ is a clutter on E. Clearly $d(p(\mathfrak{R}))$ is the $\mathcal{S} = b(\mathfrak{R})$ specified in the theorem.

Property (3) for clutters $\mathfrak{R}$ and $\mathcal{S} = d(p(\mathfrak{R}))$ is equivalent to the obvious fact that:

(4) *Every subset E_0 of E either contains a member of $\mathfrak{R}$ or is contained in a member of $p(\mathfrak{R}) = d(\mathcal{S})$, but not both.*

The equivalence follows because E_0 is contained in a member of $d(\mathcal{S})$ if and only if $E_1 = E - E_0$ contains a member of $\mathcal{S}$.

We must verify that $\mathcal{S} = d(p(\mathfrak{R}))$ is the only clutter on E such that (3) holds for $(\mathfrak{R}, \mathcal{S})$. This follows because $p(\mathfrak{R})$, as defined, is the only clutter on E for which (4) holds. To see this, suppose that clutter $\mathcal{P}$, in place of $p(\mathfrak{R})$, satisfies (4). For any $P \in \mathcal{P}$, P cannot contain a member of $\mathfrak{R}$ since P is contained in a member of $\mathcal{P}$ (itself). Because $\mathcal{P}$ is a clutter, any set $A \subset E$ which properly contains P is not contained in any member of $\mathcal{P}$. Therefore, by (4), any such A contains a member of $\mathfrak{R}$. Thus P is a maximal subset of E containing no member of $\mathfrak{R}$, and thus we conclude that $\mathcal{P} \subset p(\mathfrak{R})$. On the other hand, for any $Q \in p(\mathfrak{R})$, Q is not properly contained in any member of $\mathcal{P}$ since $p(\mathfrak{R})$ is a clutter and since $\mathcal{P} \subset p(\mathfrak{R})$. Therefore we have $Q \in \mathcal{P}$, since otherwise $E_0 = Q$ would be a set which contains no member of $\mathfrak{R}$ and which is contained in no member of $\mathcal{P}$. Thus we conclude that $\mathcal{P} = p(\mathfrak{R})$. This completes the proof of Lemma B.

Proof of Lemma C: Suppose that $(\mathfrak{R}, \mathcal{S})$ is any pair of families on E that satisfies property (3), and let f be any real-valued function on E. We shall show that equation (1) holds, i.e., that $(\mathfrak{R}, \mathcal{S})$ is a blocking system.

To compute $u(\Re, f)$, we use the following "threshold method." It is different from previously proposed algorithms for special bottleneck problems.

Choose elements $x \in E$ one after another in order of non-decreasing magnitude of $f(x)$ until the set of chosen elements first contains an $R \in \Re$. When this happens, stop. Denote the final set of chosen elements by B_u, denote the last chosen element by x_u, and denote any one of the members of $\Re$ contained in B_u by R_u (there may be several). We have $x_u \in R_u$ since $B_u - x_u$ contains no $R \in \Re$. Element x_u maximizes f over B_u and thus over R_u. Therefore $u(\Re, f) \leqslant f(x_u)$. Since $B_u - x_u$ contains every x such that $f(x) < f(x_u)$, if there were an $R \in \Re$ such that

$$\max_{x \in R} f(x) < f(x_u),$$

we would have $R \subset B_u - x_u$. Therefore $u(\Re, f) = f(x_u)$.

By property (3), $B_w = E - (B_u - x_u)$ contains a member S_w of $\math8{S}$. By property (3), $B_w - x_u = E - B_u$ contains no member of $\math8{S}$, and so we have $x_u \in S_w$. Element x_u minimizes f over B_w and thus over S_w. Therefore $f(x_u) \leqslant w(\math8{S}, f)$. Since $B_w - x_u$ contains every x such that $f(x_u) < f(x)$, if there were an $S \in \math8{S}$ such that

$$f(x_u) < \min_{x \in S} f(x),$$

we would have $S \subset B_w - x_u$. Therefore $f(x_u) = w(\math8{S}, f)$. This completes the proof of Lemma C and the bottleneck theorem.

One can of course use the "dual threshold method" instead. That is, choose elements $x \in E$ one after another in order of non-increasing magnitude of $f(x)$ until the set of chosen elements first contains an $S \in \math8{S}$.

The concept of a blocking system of clutters $\Re$ and $\math8{S}$ or of the blocking system of families $\Re^+$ and $\math8{S}^+$, where $\Re^+$ and $\math8{S}^+$ consist of all supersets of members of $\Re$ and $\math8{S}$, respectively, arises in other contexts besides bottleneck extrema (see [8], [10], [11], [13], for example). In particular, the families $\Re^+$ and $\math8{S}^+$ are Boolean duals of each other (the Boolean dual $\mathfrak{F}^*$ of a family $\mathfrak{F}$ on E consists of those subsets $H \subset E$ such that $E - H$ is not a member of $\mathfrak{F}$).

3. SOME EXAMPLES OF BLOCKING SYSTEMS

A *transversal* of an n by n array M is a subset of the positions in M such that there is exactly one member of the subset in each line of M. (A *line* of an array is either a row or a column of the array.) If clutter S consists of the transversals of M, its blocker $\mathfrak{R}$ consists of the h by k subarrays of M with $h + k = n + 1$. This is the blocking system for the bottleneck assignment problem.

As stated earlier, if S consists of the arc-sets of paths from node a to node b in a graph G (perhaps directed), then the members of its blocker $\mathfrak{R}$ are called the cuts separating b from a.

If clutter $\mathfrak{R}$ consists of the arc-sets that are complementary to spanning trees in a graph G, then S consists of the arc-sets of circuits (polygons) in G.

If $\mathfrak{R}$ consists of the minimal sets of nodes that meet all arcs in a graph G, then S consists of the pairs of adjacent nodes in G.

In each of these examples, there is a good algorithm for recognizing whether a given subset E_0 of the domain E contains a member of the clutter $\mathfrak{R}$ or whether its complement $E_1 = E - E_0$ contains a member of clutter S.

Very often it is difficult to find a useful description of the blocking clutter of a simply described clutter, and very often it is difficult to evaluate a bottleneck extremum. In view of the threshold method for bottleneck extrema, it is clear that having a good algorithm for a bottleneck problem, defined by any clutter $\mathfrak{R}$ of some class of clutters and by any function f on the domain E of $\mathfrak{R}$, is equivalent to having a good algorithm for determining, for any $\mathfrak{R}$ of the class and any subset $E_0 \subset E$, whether or not E_0 contains a member of $\mathfrak{R}$, i.e., for determining whether E_0 contains a member of $\mathfrak{R}$ or whether $E - E_0$ contains a member of $b(\mathfrak{R})$. A necessary, though not sufficient, condition for the latter is having a good algorithm for recognizing whether any given subset of E is itself a member of $\mathfrak{R}$ or a member of $b(\mathfrak{R})$. For any clutter $\mathfrak{R}$ of direct interest, it is likely that its members are easily recognizable. Unfortunately, this does not imply that the same is true for $b(\mathfrak{R})$.

The theorems below may be interpreted as describing good algorithms for recognizing members of the blocking clutters of certain clutters $\mathcal{R}$. Good algorithms are known also (though we will not describe them here) for determining, for any one of these particular clutters and any subset of its domain, whether or not the subset contains a member of the clutter.

The description of the blocking clutter of the clutter of transversals in a square array is, in spite of its simple appearance, a quite substantial theorem. In view of property (3), it is clearly equivalent to the following: For any n by n array M and any subset E_0 of positions in M, E_0 contains no transversal of M if and only if there are $2n - (n + 1) = n - 1$ lines of the array that together contain all of E_0. This is a special case of the well-known König theorem: For any rectangular array M and any subset E_0 of its positions, the maximum cardinality of a matching contained in E_0 equals the minimum number of lines that together contain all of E_0. (A *matching* is a set of positions, no two of which lie in the same line.) The König theorem is equivalent to the following description of a more general class of blocking systems: If clutter $\mathcal{R}$ consists of the matchings of cardinality t in an m by n array, then $b(\mathcal{R})$ consists of all h by k subarrays such that $h + k = m + n - t + 1$. A good algorithm for determining whether a given subset of the positions in an m by n array contains a matching of size t is described in [3].

Another blocking system based on m by n arrays can be obtained from the linear programming transportation problem: Let $X = (x_{ij})$ be an extreme solution (basic feasible solution) of the constraints

$$\sum_{j=1}^{n} x_{ij} = r_i, i = 1, \cdots, m, \quad \sum_{i=1}^{m} x_{ij} = s_j, j = 1, \cdots, n, \qquad x_{ij} \geqslant 0,$$

where r_i and s_j are given non-negative numbers satisfying

$$\sum_{i=1}^{m} r_i = \sum_{j=1}^{n} s_j.$$

The *support* of X is the subset of positions (i, j) such that $x_{ij} > 0$. Then the family of all supports of extreme solutions X is a clutter $\mathcal{R}$ on the domain E of positions (i, j), and $b(\mathcal{R})$ consists of all minimal subarrays $I \times J$ (where $I \subset \{1, 2, \cdots, m\}, J \subset \{1, 2, \cdots, n\}$) such that

$$\sum_{i \in I} r_i + \sum_{j \in J} s_j > \sum_{i=1}^{m} r_i.$$

This description of $b(\mathfrak{R})$ can be deduced from the max-flow min-cut theorem of Ford and Fulkerson [3]. Here the bottleneck problem, evaluate $u(\mathfrak{R}, f)$ for any given real-valued function f on the set E of positions (i, j), has the interpretation: Satisfy all the "demands" s_j from the "supplies" r_i in the least time, f_{ij} being the transportation time from supply point i to demand point j. There are good network-flow algorithms for determining whether a given subset of positions contains the support of a solution X.

Let E consist of all the unordered pairs of objects in a finite set V. A *perfect matching* of V is a subset of E whose members are disjoint and together contain all of V. Let clutter $\mathfrak{R}$ consist of all the perfect matchings of V. Then $\mathcal{S} = b(\mathfrak{R})$ consists of the subsets $S(\mathcal{P})$ of E obtained as follows: $\mathcal{P}$ is any family of mutually disjoint, odd-cardinality subsets of V such that $|V| - |\cup(\mathcal{P})| = |\mathcal{P}| - 2$; $x \in E$ is a member of $S(\mathcal{P})$ if and only if the two members of x are members of different members of $\mathcal{P}$. This result is equivalent to Tutte's theorem characterizing those subsets E_0 of E that contain a perfect matching [14]. (See [1].) Edmonds [2] has given a good algorithm for determining whether a subset E_0 of E contains a perfect matching.

One of the many classes of clutters $\mathfrak{R}$ for which $b(\mathfrak{R})$ is generally a mystery is where $\mathfrak{R}$ consists of the arc-sets of Hamilton tours in a graph. The bottleneck traveling salesman problem, like the traveling salesman problem, is also a mystery. There is no known good algorithm for determining whether a given subset of the arcs of a graph contains a member of $\mathfrak{R}$.

REFERENCES

1. Berge, C., *Théorie des Graphes et ses Applications*, Dunod, Paris, 1958.
2. Edmonds, J., "Paths, Trees, and Flowers," *Canad. J. Math.* **17** (1965), 447–467.

3. Ford, Jr., L. R., and D. R. Fulkerson, *Flows in Networks*, Princeton University Press, Princeton, N.J., 1962.
4. Fulkerson, D. R., "Flow Networks and Combinatorial Operations Research," *Amer. Math. Monthly* **73** (1966), 115–138.
5. Fulkerson, D. R., I. Glicksberg, and O. Gross, *A Production Line Assignment Problem*, The RAND Corporation, RM-1102, 1953.
6. Gilmore, P. C., and R. E. Gomory, "Sequencing a One State-Variable Machine: A Solvable Case of the Traveling Salesman Problem," *Operations Res.* **12** (1964), 655–679.
7. Gross, O., *The Bottleneck Assignment Problem*, The RAND Corporation, P-1630, 1959.
8. Hu, S. T., *Threshold Logic*, University of California Press, Berkeley, California, 1965.
9. Hu, T. C., "The Maximum Capacity Route Problem," *Operations Res.* **9** (1961), 898–900.
10. Lawler, E., "Covering Problems: Duality Relations and a New Method of Solution," *SIAM J.* **14** (1966), 1115–1133.
11. Lehman, A., "On the Width Length Inequality," to appear in *SIAM J.*
12. Pollack, H., "The Maximum Capacity Route through a Network," *Operations Res.* **9** (1960), 722–736.
13. Shapley, L. S., "Simple Games: An Outline of the Descriptive Theory," *Behavioral Sci.* **7** (1962), 59–66.
14. Tutte, W. T., "The Factors of Graphs," *Canad. J. Math.* **4** (1952), 314–329.

ON CORES AND INDIVISIBILITY*

Lloyd Shapley

and

Herbert Scarf

1. INTRODUCTION

This paper has two purposes. To the reader interested in the mathematics of optimization, it offers an elementary introduction to n-person games, balanced sets, and the core, applying them to a simple but nontrivial trading model. To the reader interested in economics, it offers what may be a new way of looking at the difficulties that afflict the smooth functioning of an economy in the presence of commodities that come in large discrete units.

The core of an economic model, or of any multilateral competitive situation, may be described as the set of outcomes that are

* This research was supported by National Science Foundation grant GS-31253. The authors are indebted to Professors David Gale and Bezalel Peleg for several helpful comments.

"coalition optimal," in the sense that they cannot be profitably upset by the collusive action of any subset of the participants, acting by themselves. There is no reason, *a priori*, that such outcomes must exist; the core may well be empty. But it has been shown that important classes of economic models do have nonempty cores. In fact, whenever a system of *competitive prices* exists (prices under which individual optimization decisions will lead to a balance of supply and demand), the resulting outcome is in the core.* The core may also exist in the absence of competitive prices. It is of some interest, therefore, to relax one or more of the classical "regularity" assumptions that, taken together, ensure the existence of competitive prices—such as convexity of preferences, perfect divisibility of commodities, constant returns to scale in production, and absence of externalities—and to ask under what conditions the resulting system will have a core. There is already a considerable literature in this area.**

We stress that the core is a general game-theoretic concept, definable without reference to any market model.*** Moreover, its existence for the classical exchange economies can be established, if we wish, without making use of the idea of competitive prices.****

In this paper we consider the case of a commodity that is inherently indivisible, like a house. We formulate a class of markets in which a consumer never wants more than one item, but has ordinal preferences among the items available. We then prove that a core always exists for this model, making use of the concept of "balanced sets." The competitive prices for this model are next determined by a separate argument, providing an alternative and more constructive proof of the existence.

A counterexample is then considered, showing that if more complex schemes of preferences among the indivisible goods are al-

---

* See [5].

** For nonconvexity, see [12, 19, 25]; for indivisibility, see [11, 12, 22, 24]; for nonconstant returns, see [14, 16]; for externalities, see [7, 21]. Not all of these refer directly to the core.

*** See [1, 3, 15, 17, 18].

**** See [15, 18, 20].

lowed, the core may disappear, even though all of the classical conditions except perfect divisibility are satisfied.

In the final section, for perspective, we review a series of other models involving indivisible commodities that have been discussed in the literature from the viewpoint of the core.

2. THE MODEL

Let there be n traders in the market, each with an indivisible good to offer in trade (e.g., a house). The goods are freely transferable, but we shall assume that a trader never has use for more than one item. There being no money or other medium of exchange, the only effect of the market activity is to redistribute the ownership of the indivisible goods, in accordance with the (purely ordinal) preferences of the traders. We shall describe these preferences with the aid of a square matrix: $A = (a_{ij})$, where $a_{ij} > a_{ik}$ means that trader i prefers item j to item k, and $a_{ij} = a_{ik}$ means that he is indifferent.* Owning no items, we assume, is ranked below all else, and owning several items is ranked only equal to the maximum of their separate ranks. Although only ordinal comparisons are involved in the model, it will be convenient to think of A as a matrix of real numbers.

The final effect of any sequence of transfers can be described by another matrix $P = (p_{ij})$, called an *allocation*, in which $p_{ij} = 1$ if trader i holds item j at the end of trading, and $p_{ij} = 0$ otherwise. In the interesting cases, P will be a permutation matrix, i.e., a zero-one matrix with row-sums and column-sums all equal to 1. In any case, the column-sums of P will be equal to 1.

Let N denote the set of all traders, and let $S \subseteq N$. By an S-*allocation* P_S, we shall mean an n-by-n zero-one matrix containing one 1 in each column indexed by a member of S, but containing only zeros in the rows and columns indexed by members of $N - S$. An S-allocation describes a distribution of goods that the "submarket" S could effect. An S-allocation with no row-sum greater

---

* By "item j" we mean the good brought to the market by trader j.

than 1 is called an *S-permutation*. It is clear that for every S-allocation that is not an S-permutation, there is an S-permutation that is at least as desirable to every member of S, and more desirable to at least one member of S.

An allocation will be said to be a *core allocation* if there is no submarket S that could have done better for all its members. A core allocation, therefore, is one that cannot be improved upon by "recontracting" in the sense of Edgeworth [6]. Our aim is to show that *every market of the kind described possesses at least one core allocation*.

3. GAMES AND CORES

First, let us recast the problem in a game-theoretic form. Let E^N denote the n-dimensional Euclidean space with coordinates indexed by the elements of the set N, and similarly E^S for $S \subset N$. For $x, y \in E^N$ and $S \subseteq N$, we define $y \geq_S x$ to mean that $y_i \geq x_i$, all $i \in S$. The notations $>_S$ and $=_S$ are defined similarly.

A "cooperative game without side payments" [1, 3, 15, 18] will be identified with its "characteristic function." This is a function V from the nonempty subsets of the "player space," N, to the subsets of the "payoff space," E^N, satisfying the following conditions for each S, $\emptyset \subset S \subseteq N$:

(a) $V(S)$ is closed.

(b) If $x \in V(S)$ and $x \geq_S y$, then $y \in V(S)$.

(c) $[V(S) - \cup_{i \in S} \text{int } V(\{i\})] \cap E^S$ is bounded and nonempty.

Here "int" denotes "interior of."* Note that property (b) implies that each $V(S)$ is a *cylinder* (i.e., the Cartesian product of E^{N-S} with a subset of E^S). Intuitively, the projection of $V(S)$ on E^S is supposed to represent the payoffs that the members of S, acting cooperatively, can achieve (or exceed) without outside help.

This interpretation suggests a fourth property, namely, *super-*

---

* Condition (c) implies, in particular, that none of the $V(S)$ is either empty or equal to E^N.

additivity, which may be expressed as

(d) $V(S \cup T) \supseteq V(S) \cap V(T)$ if $S \cap T = \emptyset$.

Although this condition often holds in practice, it will not be required in the definition of a game.

Of particular interest is a class of games in which the $V(S)$ are generated by *finite* sets $Y(S) \subseteq E^N$, as follows:

$$V(S) = \{x : y \geqq_S x \text{ for some } y \in Y(S)\}.$$

Assuming that there are no superfluous generators, each payoff $y \in Y(S)$ identifies what might be called a "corner" of the cylindrical set $V(S)$.* Finitely generated games arise often in applications; they also figure prominently in the proof of the main theorem of [15].**

The *core* of the game may be defined as the set

$$V(S) - \bigcup_{\emptyset \subset S \subset N} \text{int } V(S).$$

In other words, the core is the intersection of $V(N)$ with the closures of the complements of *all* of the $V(S)$, including $V(N)$ itself. The core is a closed subset of the boundary of $V(N)$, possibly empty but certainly bounded, and every point in the core is (weakly) Pareto optimal.

Intuitively, the core consists of those outcomes of the game that are *feasible* (i.e., are in $V(N)$), and that cannot be *improved upon* by any individual or coalition of individuals (i.e., are not interior to any $V(S)$). It is of considerable interest in the analysis of any cooperative game to determine whether its core is nonempty.

Finally, we need the notion of a "balanced" game. Let us call a family T of nonempty subsets of N *balanced* if the system of equa-

---

* That is, a vertex of the projection of $V(S)$ on E^S.

A glance ahead to Figs. 1, 2, or 4 may aid the reader in visualizing the definitions of this section.

** But not in the alternative proof given in [18].

tions

$$\sum_{S\,:\,j \in S} \delta_S = 1, \qquad j \in N,$$

has a nonnegative solution with $\delta_S = 0$ for all S not in T.* A partition of N is a simple example of a balanced family. The numbers δ_S are called *balancing weights* for T; it has been shown [12] that they are unique for T if and only if no proper subfamily of T is balanced.

The game V is said to be *balanced* if the following inclusion statement:

(e) $\bigcap_{S \in T} V(S) \subseteq V(N)$,

holds for *all* balanced families T. Property (e) is obviously related to property (d) via the partitions of N, but neither condition directly implies the other. A fundamental theorem states that the core of a balanced game is not empty.** We shall apply this theorem to the market described in Sec. 2.

4. THE CORE OF THE MARKET

We now return to the market model of Sec. 2. First, we must determine the characteristic function. The sets $V(S)$ are finitely generated by the S-permutations, since all other S-allocations are dominated by S-permutations in the sense of (b) above. It will be convenient to express $V(S)$ with the aid of a zero-one matrix $B_S(x)$, defined for each $x \in E^N$ as follows:

$$b_{S|ij}(x) = \begin{cases} 1 & \text{if} \quad a_{ij} \geqq x_i \quad \text{and} \quad i \in S, \\ 0 & \text{if} \quad a_{ij} < x_i \quad \text{or} \quad i \notin S. \end{cases}$$

Thus, $B_S(x)$ tells us, for each trader i in S, exactly which items he ranks *at or above* his utility level x_i. We can then define the

---

* See [10, 13, 17, 18].

** See [4, 15, 17, 18] or the survey article [3].

game V as follows:

$$V(S) = \{x : B_S(x) \geqq P_S \text{ for some } S\text{-permutation } P_S\}.$$

THEOREM: *V is a balanced game; hence the market in question has a nonempty core.*

Proof: It is immediately evident that V satisfies conditions (a), (b), and (c) of Sec. 3. (We could also show without difficulty that V satisfies (d).) It remains to be shown that V satisfies (e).

Let T be any balanced family of coalitions, and let $x \in \bigcap_{S \in T} V(S)$. Let $\{\delta_S\}$ be balancing weights for T. Then we have

$$B_N(x) = \sum_{S \in T} \delta_S B_S(x).$$

By definition of $V(S)$, there exists an S-permutation P_S, for each $S \in T$, such that $B_S(x) \geqq P_S$, and so

$$\sum_{S \in T} \delta_S B_S(x) \geqq \sum_{S \in T} \delta_S P_S.$$

Call the matrix on the right D; then we have

$$B_N(x) \geqq D.$$

The crucial fact about D is that it is doubly stochastic; that is, it is nonnegative and has all row- and column-sums equal to 1. This follows directly from the definition of balancing weights; thus, the ith row sum is

$$\sum_{j \in N} \sum_{S \in T} \delta_S p_{S|ij} = \sum_{S \in T} \delta_S \sum_{j \in N} p_{S|ij}$$

$$= \sum_{S \in T} \delta_S \begin{cases} 1 & \text{if } i \in S \\ 0 & \text{if } i \notin S \end{cases} = \sum_{S:i \in S} \delta_S = 1,$$

and the argument for the column sums is the same.

The next step will be to change D into a permutation matrix P_N, that is, to eliminate any fractional entries without changing the row or column sums—and to do so without disturbing any entries that are already 0 or 1. Since all entries of $B_N(x)$ are 0 or 1, we will thereby ensure that $B_N(x) \geqq P_N$.

Since a fraction cannot occur alone in a row or column, either D is already a permutation matrix or there is a closed loop of fractional entries:

Alternately adding and subtracting a fixed number ϵ to the elements of this loop will clearly preserve row and column sums. If ϵ is too large, then negative entries will be created, but making ϵ as large as possible consistent with nonnegativity will produce a new doubly stochastic matrix D' that has at least one more zero than D, and hence *fewer* fractional entries. If D' is not yet doubly stochastic, we can repeat the operation. Eventually we must obtain what we want—a permutation matrix P_N such that $B_N(x) \geqq P_N$. Hence x is in $V(N)$. Hence $\bigcap_{S \in T} V(S) \subseteq V(N)$. Hence the game is balanced.   Q.E.D.

5. AN EXAMPLE TO ILLUSTRATE THE THEOREM

Let $n = 3$, and let

$$A = \begin{pmatrix} 0 & 1 & 0 \\ 1 & 0 & 1 \\ 0 & 1 & 0 \end{pmatrix}.$$

This means that the first and third traders want only item 2, while the second trader wants either item 1 or item 3, indifferently.

The characteristic function, being finitely generated, can be described in terms of its "corners" as follows:*

$$
\begin{array}{ll}
V(\{1\}): & (0, -, -), \\
V(\{2\}): & (-, 0, -), \\
V(\{3\}): & (-, -, 0), \\
V(\{1, 2\}): & (1, 1, -), \\
V(\{1, 3\}): & (0, -, 0), \\
V(\{2, 3\}): & (-, 1, 1), \\
V(\{1, 2, 3\}): & (1, 1, 0) \text{ and } (0, 1, 1).
\end{array}
$$

As shown in Fig. 1, this game has an L-shaped core, with successive vertices $(1, 1, 0)$, $(0, 1, 0)$, $(0, 1, 1)$.

It is a curious fact that *all* outcomes in the core of this example are "weakly" improvable, in the sense that one member of an effective coalition can do better while the other member does no worse. The reader may verify that $\{2, 3\}$ can weakly improve upon any point in the core except $(0, 1, 1)$, while $\{1, 2\}$ can weakly improve upon that point as well as any other point in the core except $(1, 1, 0)$. This illustrates the fact that the "strict" improvability implicit in the definition of the core cannot be dispensed with, unless we are willing to give up the existence theorem for balanced games.

It should not be overlooked that what we are calling "outcomes" of the game do not always correspond to actual trades in the market. Indeed, there are only finitely many ways in which the goods can be reallocated. Only if one allowed some sort of free disposability, permitting the traders to diminish at will the value of the goods,** would it be possible to realize an arbitrary payoff

---

* See Sec. 3. Here, for example, $(1, 1, -)$ means the set $\{(1, 1, x_3): x_3$ arbitrary$\}$. Note that since $V(N)$ happens to require more than one generator in this example it is a nonconvex set.

** The reader with some experience in the paradoxes of bargaining will recognize that it is not inconceivable that deliberately damaging one's goods might change the overall pattern of trade in a way that returns an advantage to the damager.

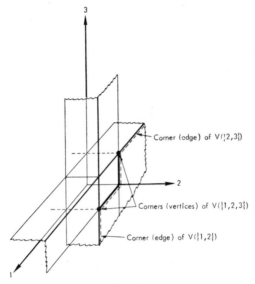

Fig. 1

vector in the "feasible" set $V(N)$. This observation does not affect the *existence* of core allocations in the market when a core exists for the game, since every "unrealizable" point in $V(N)$ is majorized by a "realizable" point. In the present example, of course, only the two tips of the "L" represent actual trades in the market.

6. COMPETITIVE PRICES

After the proof in Sec. 4 had been discovered, David Gale pointed out to the authors a simple constructive method for finding competitive prices in this market, and hence a point in its core. The following is based on his idea.

Let $R \subseteq N$, and define a *top trading cycle for R* to be any set S, $\emptyset \subset S \subseteq R$, whose s members can be indexed in a cyclic order:

$$S = \{i_1, i_2, \cdots, i_s = i_0\},$$

in such a way that each trader i_ν likes the $i_{\nu+1}$'st good at least as well as any other good in R. It is evident that every nonempty $R \subseteq N$ has at least one top trading cycle, for we may start with any trader in R and construct a chain of best-liked goods that eventually must come back to some earlier element.*

Using this idea, we can partition N into a sequence of one or more disjoint sets:

$$N = S^1 \cup S^2 \cup \cdots \cup S^p,$$

by taking S^1 to be any top trading cycle for N, then taking S^2 to be any top trading cycle for $N - S^1$, then taking S^3 to be any top trading cycle for $N - (S^1 \cup S^2)$, and so on until N has been exhausted. We can now construct a payoff vector x by carrying out the indicated trades within each cycle. That is, if $i = i_\nu^j \in S^j$, then x_i is i's utility for the good of trader $i_{\nu+1}^j$. We assert (1) that x, so constructed, is in the core, and (2) that a set of competitive prices exists for x.

To establish (1), let S be any coalition. Consider the first j such that $S \cap S^j \neq \emptyset$. Then we have

$$S \subseteq S^j \cup S^{j+1} \cup \cdots \cup S^p = N - (S^1 \cup \cdots \cup S^{j-1}).$$

Let $i \in S \cap S^j$. Then i is already getting in x the highest possible payoff available to him in S. No improvement is possible for him, unless he deals outside of S. Hence S cannot strictly improve, and it follows that x is in the core.

To establish (2), we merely assign arbitrary prices

$$\pi^1 > \pi^2 > \cdots > \pi^p > 0$$

to the goods belonging to the respective cycles S^1, S^2, $\cdots$, S^p. Then trader i in S^j can sell his own item for π^j "dollars." He cannot afford any items from S^1, $\cdots$, S^{j-1}, and so his utility is maximized if he buys the item of his cyclic successor in S^j. This, of course, costs him precisely π^j "dollars" and yields him the payoff x_i.

Not surprisingly, given the discreteness of the model, the condi-

---

* A top trading cycle may consist of a single trader!

tions on the prices are purely ordinal. To the extent that there may be different ways of constructing top trading cycles, nonuniqueness may occur in the final outcome as well as in the price ordering. It is easily seen, however, that there are no other competitive prices beyond those constructed in the above fashion, except that when two or more *disjoint* top trading cycles exist, at any stage of the construction, they may be assigned equal prices.

In the example of Sec. 5, either $\{1, 2\}$ or $\{2, 3\}$ will serve as the first top trading cycle S^1, so that we can use either $\pi_1 = \pi_2 > \pi_3$ or $\pi_3 = \pi_2 > \pi_1$ for the competitive prices. The corresponding competitive outcomes are the two tips of the L-shaped core (Fig. 1). Note that $\pi_1 = \pi_2 = \pi_3$ would *not* be competitive, as the demand for item 2 would then exceed the supply. Thus, we see from this example that the set of competitive prices is not necessarily closed.

7. ANOTHER EXAMPLE

It may be wondered in models of this type whether the core is really more general than the competitive solution. In the previous example the core of the *game* contained many "outcomes" in addition to the competitive outcome, but none of them was realizable in the *market*, given the indivisibility and undamageability of the goods. One might suspect that every *core allocation* is necessarily competitive, after all.

A counterexample is provided, however, by the following preference matrix:

goods

	0	1	2
traders	1	0	−1
	−1	1	0

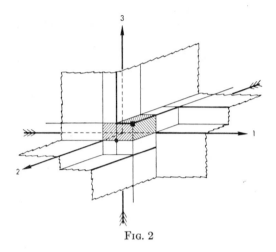

Fig. 2

The characteristic function is depicted in Fig. 2; the point marked ■ is the unique "corner" of $V(N)$. Here there is a unique top trading cycle, namely N itself, so the competitive prices are all equal and the unique competitive payoff is $(2, 1, 1)$, resulting from the N-permutation

$$\begin{pmatrix} 0 & 0 & 1 \\ 1 & 0 & 0 \\ 0 & 1 & 0 \end{pmatrix}.$$

This is of course the point ■. The core of the game evidently consists of the three shaded rectangles grouped around that point, and we see that it contains the point ● = $(1, 1, 0)$ that results from the N-permutation

$$\begin{pmatrix} 0 & 1 & 0 \\ 1 & 0 & 0 \\ 0 & 0 & 1 \end{pmatrix},$$

which is *not* a competitive allocation.

This example is not completely satisfying, since the noncompetitive outcome ● in the core is weakly majorized by the competitive outcome ■. Perhaps a better example exists, with more traders. However, as we have already remarked, we cannot base the definition of the core on weak majorization without losing the fundamental existence theorem.

8. MORE COMPLEX PREFERENCES: A COUNTEREXAMPLE

Let three traders have symmetric holdings in a tract of nine houses, as shown in Fig. 3. (Thus, trader 1 owns houses 1, 1', and 1''.) For reasons inscrutable, each trader wants to acquire three houses in a row, including exactly one of his original set. Moreover, each prefers the long row that meets this condition to the short row.

We shall show that this example of a slightly more general trading game than the preceding is not balanced, and indeed has no core—thereby dispelling any idea that the core might prove a universal remedy for market failure due to indivisibility.

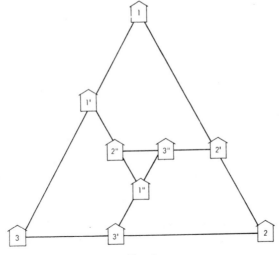

FIG. 3

The configuration of the tract is such that any two traders can make a profitable exchange. For example, a swap of 1′ and 1″ for 2 and 2′ gives trader 1 his long row and trader 2 his short row. Let us assign numerical values 2, 1, and 0 to the possession of the long row, the short row without the long row, and neither row, respectively. Then the two-person coalitions have single "corners":

$$V(\{1, 2\}): \quad (2, 1, -),$$
$$V(\{1, 3\}): \quad (1, -, 2),$$
$$V(\{2, 3\}): \quad (-, 2, 1),$$

as shown in Fig. 4. The three-person coalition cannot improve upon these pairwise exchanges; so its characteristic function is generated

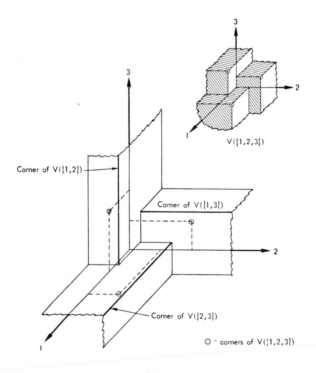

Fig. 4

by three "corners":

$$V(\{1, 2, 3\}): \qquad \begin{cases} (2, 1, 0), \\ (1, 0, 2), \\ (0, 2, 1), \end{cases}$$

as shown in the inset. Of course, the three singleton coalitions can achieve only 0.

It is easy to see that this game is *not* balanced (condition (e) in Sec. 3), since the point $(1, 1, 1)$ is in $V(\{1, 2\}) \cap V(\{1, 3\}) \cap V(\{2, 3\})$, but not $V(\{1, 2, 3\})$. It is equally easy to see that the game has no core,* since each of the generators of $V(\{1, 2, 3\})$ is interior to one of the $V(\{i, j\})$. In other words, the set $V(\{1, 2, 3\})$ is completely hidden by the union of the $V(\{i, j\})$, as suggested by the broken lines in the main figure.

It might be asked whether the absence of a core in this example might not arise from some intrinsic nonconvexity in the preference sets themselves that has nothing to do with indivisibility. The answer is no; to see this, define

$$U^1(x) = \min[2, \min(x_1, x_2, x_{2'}) + 2 \min(x_{3''}, x_{1''}, x_{3'})],$$

with $U^2(x)$ and $U^3(x)$ similar. These utility functions are concave in the nine real variables $x_1, \cdots, x_{3''}$ (since taking a "min" preserves concavity), and so generate convex preference sets. Moreover, if we restrict the variables to be 0 or 1 we obtain just the payoffs used above. Thus, the actual preference sets are nonconvex *only* because of indivisibility.

9. OTHER INDIVISIBILITY MODELS

Several other trading models with indivisible goods have been considered in the literature from the point of view of the core. They range from fairly general situations for which positive results can be obtained to specific illustrative counterexamples.

---

* We remind the reader that balancedness is sufficient, but not necessary, for a nonempty core.

In the "marriage market" or "dance floor" of [9], there are two *types* of traders. The members of each type rank those of the other type in order of preference as partners; then they pair off. There are generally many core allocations for this situation, that is, arrangements into pairs such that no two individuals of opposite type could do better. Curiously enough, it may be that none of these core allocations gives anyone his (or her) first choice. (In contrast, the "top trading cycle" construction of Sec. 6 obviously ensures that at least one trader gets his first choice.) A simple "courtship" algorithm is described in [9] for reaching a point in the core.* In fact, two extremal core allocations are reached: one gives every member of the first type the best outcome possible within the core; the other does the same for the second type. These two allocations coincide only in the case of a one-point core.**

It does not appear to be possible to set up a conventional market for this model, in such a way that a competitive price equilibrium will exist and lead to an allocation in the core.

A very similar model is the "problem of the roommates," also discussed in [9]. The difference is that there are no types; each trader ranks *all* the others as potential partners before they pair off. In this case, a very simple four-person example shows that there need not be a core.***

---

   * The deferred acceptance algorithm of [9] is easily described: (1) Let each boy propose to his favorite girl; (2) let each proposed-to girl keep her best suitor waiting and reject all others; (3) let each rejected boy propose to his next-best girl; (4) repeat steps (2) and (3) until either there are no rejected boys left or every rejected boy has exhausted the list of girls. The resulting pairing-off corresponds to a point in the core, and no boy can do better at any other point in the core.

   See also *The New Yorker*, Sept. 11, 1971, p. 94.

   ** Similar results hold for the more general "college admissions market," in which each trader of the first type can accommodate a large number of traders of the second type; see [9].

   *** Let A rank $B > C > D$; let B rank $C > A > D$; let C rank $A > B > D$; and let D rank arbitrarily. Then no pairing is stable, in the sense of the core. For example, $(AB)(CD)$ can be improved upon by the coalition $\{B, C\}$.

In the "treasure hunt" of [23],* a party of explorers finds a big cache of treasure chests in the desert, momentarily exposed during a sandstorm, but so heavy that it takes two men to carry each chest to high ground. If we consider the bargaining game that determines how the profits are to be divided, it is not difficult to see that the existence of a core depends on whether the size of the party is even or odd. Similarly, in the "Bridge game" economy of [24], an exact multiple of four players is required, if the card party is to have a core.

In most of these examples, the real issue is the indivisibility of the participants themselves, rather than the indivisibility of some more or less tangible economic commodity that is owned and is transferable. The individual is required to participate fully and exclusively in a single activity in order to have any effect. Thus, these examples could also be regarded as instances of increasing returns to scale in the labor inputs to certain production processes, or even as nonconvexities in the preferences for certain forms of consumption.**

In the "assignment game" of [22],*** there are again two types of traders, namely, *sellers* and *buyers*. The first have houses, say, and the second have money. Preferences are not merely ordinal, but are expressed as monetary evaluations of the different houses. Since money is fully transferable and infinitely divisible, the sets $V(S)$ are not finitely generated; instead they are half-spaces, and the market reduces to a "game with side payments." Competitive prices exist and are closely related to the linear-programming solution to the problem of maximizing the total monetary value of the allocation, which turns out to be the familiar "optimal assignment" problem. These prices are usually not unique. The core in this case is exactly the set of competitive allocations, in contrast to what we

---

* Inspired by the Huston-Traven classic *Treasure of the Sierra Madre*, whose plot turns on the coalitional instability of a party of *three* prospectors in a rather similar predicament.

** Compare the "gin and tonic" example in [19], where again the existence of a core depends on the parity of the number of traders.

*** See also [8], pp. 160–162.

found in Sec. 7 above. As in the marriage market, two extremal allocations can be distinguished in the core: a "high-price" corner, which is best possible for every seller, and a "low-price" corner, which is best possible for every buyer. If the houses happen to be all alike, then the core reduces to a line segment and at any given point in the core all houses have the same price.

It would be interesting if a general framework could be found that would unify some or all of these scattered results.

REFERENCES

1. Aumann, R. J., and B. Peleg, "Von Neumann-Morgenstern solutions to cooperative games without side payments," *Bull. Am. Math. Soc.*, **66** (1960), 173–179.
2. Billera, L. J., "Some theorems on the core of an n-person game without side payments," *SIAM J. Appl. Math.*, **18** (1970), 567–579.
3. ———, "Some recent results in n-person game theory," *Math. Programming*, **1** (1971), 58–67.
4. Bondareva, O. N., "Some applications of linear programming methods to the theory of cooperative games," (Russian), *Problemy Kibernet.* **10** (1963), 119–139.
5. Debreu, G., and H. E. Scarf, "A limit theorem on the core of an economy," *Intern. Econ. Rev.*, **4** (1963), 235–246.
6. Edgeworth, F. Y., *Mathematical Psychics*, London: Kegan Paul, 1881.
7. Foley, D. K., "Lindahl's solution and the core of an economy with public goods," *Econometrica*, **38** (1970), 66–72.
8. Gale, David, *The Theory of Linear Economic Models*, New York: McGraw-Hill, 1960.
9. Gale, D., and L. S. Shapley, "College admissions and the stability of marriage," *Amer. Math. Monthly*, **69** (1962), 9–15.
10. Graver, J. E., "Maximum depth problem for indecomposable exact covers," (mimeographed, Syracuse University), paper presented at the *International Colloquium on Infinite and Finite Sets*, June 1973, Keszthely, Hungary.
11. Henry, Claude, "Indivisibilités dans une économie d'échanges," *Econometrica*, **38** (1970), 542–558.
12. ———, "Market games with indivisibilities and non-convex preferences," *Rev. Econ. Studies*, **39** (1972), 73–76.
13. Peleg, Bezalel, "An inductive method for constructing minimal balanced collections of finite sets," *Naval Res. Logist. Quart.*, **12** (1965), 155–162.
14. Rader, J. T., "Resource allocation with increasing returns to scale," *Am. Econ. Rev.*, **60** (1970), 814–825.

15. Scarf, H. E., "The core of an n-person game," *Econometrica*, **35** (1967), 50–69.
16. ———, "Notes on the core of a productive economy," unpublished.
17. Shapley, L. S., "On balanced sets and cores," *Naval Res. Logist. Quart.*, **14** (1967), 453–460.
18. ———, "On balanced games without side payments," in *Mathematical Programming* (T. C. Hu and S. M. Robinson, eds.), New York: Academic Press, 1973, pp. 261–290.
19. Shapley, L. S., and M. Shubik, "Quasi-cores in a monetary economy with nonconvex preferences," *Econometrica*, **34** (1966), 805–827.
20. ———, "On market games," *J. Econ. Theory*, **1** (1969), 9–25.
21. ———, "On the core of an economic system with externalities," *Am. Econ. Rev.*, **59** (1969), 678–684.
22. ———, "The assignment game I: The core," *Intern. J. Game Theory*, **1** (1972), 111–130.
23. ———, *Competition, Welfare, and the Theory of Games*, unpublished.
24. Shubik, Martin, "The 'Bridge Game' economy: An example of indivisibilities," *J. Pol. Econ.*, **79** (1971), 909–912.
25. Starr, R. M., "Quasi-equilibria in markets with nonconvex preferences," *Econometrica*, **37** (1969), 25–38.

MARKOV DECISION CHAINS*

*Arthur F. Veinott Jr.***

0. ABSTRACT

This is a self-contained expository development of "policy improvement" methods for finding optimal policies under various criteria in Markov decision chains. A (finite) Markov decision chain is a generalization of a finite Markov chain with a distinguished set of stopped states such that whenever the chain is observed in a given state, a decision maker chooses one of finitely many transition probability vectors available in that state and earns a reward depending on the given state and probability vector chosen. No background in the subject is required, but a knowledge of elementary properties of matrices, infinite series, and finite

---

* This work was supported by ONR contract N00014-67-A-0112-0050 and NSF grants GK-18339 and GK-13757. Reproduction in whole or in part is permitted for any purpose of the United States government.

** This paper was prepared while the author was on sabbatical leave from Stanford University at Yale University.

Markov chains is assumed. Literature citations appear at the end of the paper.

The nature of sequential decision making is discussed in Section one. This is followed in Section two with applications of Markov decision chains to problems of gambling, search, sequential statistical decisions, and inventory control. Finite policy improvement methods for constructively establishing the existence of a stationary policy maximizing an appropriate criterion are developed in Sections 3–5. Section three treats the "transient" case, i.e., where the mean time to reach the stopped states is finite from every state and under every policy. The aim is to maximize the expected reward earned before entering the stopped states. The final two sections deal with the situation where future rewards are "discounted" using a positive interest rate. The problems of finding a policy that maximizes the expected discounted reward for (i) a fixed interest rate and (ii) all small enough positive interest rates are both discussed. The key to the analysis of (ii) is the development of the expected discounted reward as a Laurent series in the interest rate. A natural way to carry out the computations is to lexicographically maximize the coefficients in the series, viz., the long run expected average earnings, the expected transient earnings, the expected temporary earnings, etc.

1. INTRODUCTION

The nature of sequential decision making.

One of the fascinating and recurrent problems of human affairs is decision making in the face of uncertainty about the possible outcomes of alternative actions. A surgeon must decide whether and when to transplant a patient's heart when availability of donors is uncertain, and both survival and death are possible outcomes in any case. An investor seeks to manage his portfolio of risky securities by striking an appropriate balance between high return and low risk, and by properly timing sales and purchases to take advantage of uncertain market conditions. An inventory manager, facing uncertain demand for a product, seeks to maintain inventory levels that avoid the excessive storage or shortage costs resulting from either over or under stocking the product.

Each of these problems is dynamic, i.e., decisions are made at several different points in time. Moreover, the actions chosen at one point in time affect the options available and environment faced at future times. Thus one must live with the consequences of one's decisions. For example, a surgeon who decides not to use a donor may find that donor is not available subsequently. The investor who fails to sell a security whose price is dropping may be forced to take a bigger loss later. The inventory manager who buys too much stock at some time must face a long interval of excessive storage costs or, perhaps, disposal at a loss.

The foregoing problems are also sequential, i.e., they are dynamic and new information is acquired as time passes. The new information acquired at a given point in time is the collection of values of the random variables which are first observed at that point and which are relevant to the choice of subsequent decisions. For example, the surgeon observes results of tests performed on his patient, the investor observes securities prices, and the inventory manager observes the actual demands for his product as they occur. In deterministic problems, i.e., where the outcomes of all actions can be predicted with certainty at the outset, no new information is acquired as time passes so the distinction between dynamic and sequential problems vanishes.

If one ignores the costs of storing and retrieving information in sequential decision problems, as we do, it is clearly best to allow the actions chosen at a given time to depend on all the relevant information acquired by that time, i.e., one should make sequential decisions. The reason is simply that one's flexibility is increased by so doing. For after all, one can always ignore some or all of the relevant information acquired, thereby making the decision depend only on the remaining information. But such a course can never improve one's performance. The essential point is that nothing is to be gained by committing oneself prematurely to a course of action.

To illustrate, the inventory manager could decide at the outset how much to purchase in each period. But such a strategy can be substantially improved upon by waiting to observe the inventory on hand at the beginning of a period (which depends on the demands and orders in all prior periods) before choosing the amount to purchase in the period. For by waiting, he can usually bring the stock level after ordering in the period much closer to the desired level in that period than would be possible if the amount ordered in the period were chosen earlier.

In practice perhaps the most important part of sequential decision making is to properly formulate a problem, to identify the alternative actions and relevant information at each point in time, and to assess the rewards and probabilities of alternate outcomes. Once this is done, there remains the often difficult problem of finding a policy that is in some sense "optimal." It is this last problem to which we address ourselves in the sequel.

The preceding informal discussion sets the stage for giving a precise formulation of sequential decision processes. We now do this for a stationary model in which there are finitely many relevant "states" of information and actions in each state. Also the total reward earned in a finite sequence of periods is the sum of the rewards in those periods.

Markov decision chains.

Consider a system which is observed at each of a sequence of points in time labeled $1, 2, \cdots$. At each of those points the system is found to be in one of S *states*, labeled (for notational simplicity only) $1, 2, \cdots, S$, or to have *stopped* by entering a finite set of *stopped states* disjoint from the set of S ordinary states. Intuitively, the state of the system at a particular point in time provides all information relevant for subsequent decisions. At each point in time at which the system is observed in state s, an *action* a is chosen from a finite set A_s of possible actions and a *reward* $r(s, a)$ is received. The conditional probability that the system is observed in state t at time $N + 1$ given that it is found in state s at time N, that action a is taken at that time, and given the observed states and actions taken at times $1, 2, \cdots, N - 1$, is assumed to be a nonnegative function $p(t \mid s, a)$ depending only on t,

s, and a. The corresponding conditional probability that the system is observed to have stopped at time $N + 1$ is $q(s, a) = 1 - \Sigma_{t=1}^{S} p(t \mid s, a) \geqq 0$. Once the system is observed to have entered a stopped state, it remains there and earns no subsequent rewards.

If the action chosen in each state at each point in time depends only on the state and the point in time, the successive states of the system form a (finite) Markov chain. For this reason, we call the above system a (finite) *Markov decision chain*.

Representation as a directed graph.

A Markov decision chain can be represented as a directed graph with labeled arcs in which the nodes are partitioned into two sets, one consisting of the set of all states (including the stopped states that are accessible from some ordinary state) and the other the set of all actions available in at least some state. There is an arc from each state to each action available in that state, the arc label being the reward earned by the given state-action combination. Moreover, there is an arc from each action (in a state) to each state to which there is a positive transition probability, the arc label being the indicated probability in this case. If an action leads to a given state with probability one, we can and often do condense the action with the state and omit the arc (and its label) leading from the action to the state. In the sequel we display graphs by using circles for states, triangles for actions, and arrows (including numbered labels) for arcs.

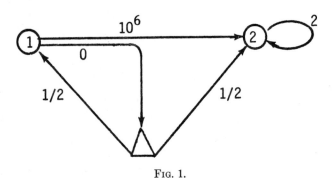

Fig. 1.

The two state example in Figure 1 illustrates these concepts. In state one, there are two actions. One action earns 10^6 and leads to state two with probability one. The other action earns zero and leads to states one and two each with probability one half. In state two there is only one action which earns two and leads back to state two with probability one.

Myopic policies.

Perhaps the simplest interesting policy that one might consider using in a Markov decision chain is a *myopic* one, i.e., a policy which chooses for each state s, an action $a = a_s \in A_s$ that maximizes the immediate reward $r(s, a)$ over $a \in A_s$. Since myopic policies ignore all but the immediate future, they cannot generally be expected to be "optimal." Nevertheless, they often are "optimal" (e.g., as in Figure 1) as the examples in Section 2 below suggest.

2. APPLICATIONS

At this point, it may be helpful to illustrate the wide scope of applications of Markov decision chains. We do so with a few examples drawn from the literature on gambling, search, sequential statistical decisions, and inventory control. In each case the aim will be to maximize the expected reward earned before entering the stopped state(s). Other criteria will be considered in Sections 4 and 5. In each problem we indicate, usually without proof, the form of the optimal policy. For further details the reader may consult the bibliographic notes. In some of these examples, we do not insist that the (finite) set of states be the first S positive integers.

Example 1. Target Problems: Gambling.

It is often desired to move a Markov decision chain from a given initial state to a target, which we take to be a distinguished stopped state t say, with maximum probability. In this event one sets $r(s, a) = p(t \mid s, a)\,(\leqq q(s, a))$.

As a simple example of this problem, consider a gambler who seeks to maximize the probability of increasing his fortune from

$s \ (= 1, \cdots, S - 1)$ dollars to at least S dollars (his target) by repeated independent play of the following game. At each play the gambler may bet any positive integer number of dollars a not exceeding his then existing fortune s, provided s is positive. The gambler wins a bet with probability p $(0 < p < 1)$ and loses it with probability $1 - p$. A win increases his fortune to $s + a$ while a loss reduces it to $s - a$. Of course, it never pays to risk more than necessary so $s + a \leq S$. Also play terminates when either the gambler goes broke or reaches his target (the two stopped states). Under these hypotheses, $A_s = \{1, \cdots, \min(s, S - s)\}$ for $0 < s < S$; $r(s, a) = p$ if $s + a = S$ and $r(s, a) = 0$ otherwise; $p(s + a \mid s, a) = p$, $p(s - a \mid s, a) = 1 - p$, and $p(t \mid s, a) = 0$ otherwise. The gambler's optimal strategy is to play *boldly*, i.e., make the largest possible bet on each play, if the game is unfavorable, i.e., if $p < \frac{1}{2}$, and play *timidly*, i.e., bet one dollar on each play, if the game is favorable, i.e., if $p > \frac{1}{2}$. If the game is fair, i.e., $p = \frac{1}{2}$, every strategy is optimal. An interesting feature of this game is that the choice between bold and timid play depends only on the "house" odds and not upon the gambler's initial fortune or his target. However, the maximum probability of eventually reaching the target does increase as the initial fortune increases and as the target is lowered. Incidentally, bold play is myopic, but timid play is not.

Example 2. Stopping Problems: Search and Sequential Bayesian Statistical Decisions.

Consider an S state Markov chain with transition probabilities p_{st}, $1 \leq s, t \leq S$, and a stopped state which is entered from state s with (a possibly zero) probability $q_s \equiv 1 - \Sigma_{t=1}^{S} p_{st}$. Suppose that upon observing the process to be in state s at some point in time, there are two actions available. One, denoted σ, involves stopping, i.e., entering the stopped state with probability one, and receiving a reward r_s. The other action, denoted γ, involves continuing, i.e., paying an entrance fee c_s to observe the process again at the following point in time. The objective is usually to maximize the total expected (net) reward. The central issue in these "stopping problems" is whether it is better to stop in a given state with a certain reward or to pay an entrance fee to continue with the uncertain prospect of stopping in a state with a higher reward later. In our notation for Markov decision chains one has for each $1 \leqq s, t \leqq S$ that $A_s = \{\sigma, \gamma\}$, $r(s, \sigma) = r_s$, $r(s, \gamma) = -c_s$, $p(t \mid s, \sigma) = 0$, $q(s, \sigma) = 1$, $p(t \mid s, \gamma) = p_{st}$, and $q(s, \gamma) = q_s$.

Many stopping problems, possibly after suitable transformation, have the properties that (i) $r_s = 0$ for all s and (ii) $p_{st} = 0$ whenever $c_s \geqq 0$ and $c_t < 0$. In these circumstances, the myopic policy, which calls for continuing in states with negative entrance fees and stopping otherwise, is optimal! To see this observe that it certainly pays to continue in states having a negative entrance fee (which amounts to a positive reward) because one can always stop with a zero reward in the following period. On the other hand, since the reward from stopping is always zero, the only reason for continuing in a state with a nonnegative entrance fee is the prospect of subsequently arriving at a state with a negative entrance fee. But this possibility is ruled out by (ii), so nothing is lost by stopping.

As an illustration of a stopping problem, and in particular one in which (i) and (ii) above hold, consider how a salvage firm might search for a sunken treasure worth $r > 0$ dollars. The firm has isolated an area of the sea in which they believe with (*a priori*) probability p, $0 < p < 1$, the treasure lies. The firm can send a diver down at a cost of c dollars per dive. The conditional probability that any particular dive is unsuccessful (i.e., no treasure is found) given that the treasure is there is q, $0 < q < 1$. The problem is to choose when to quit unsuccessful diving so as to maximize the expected net reward from the salvage operation.

A notable feature of this problem is its "statistical" character, i.e., each dive gives the firm additional information in the light of which the probability the treasure is there is revised. To see this, suppose the first s dives are unsuccessful. Then the (*a posteriori* conditional) probability the treasure is there is, by Bayes' rule,

$$p_s = \frac{pq^s}{pq^s + 1 - p}, \qquad s = 0, 1, \cdots.$$

Observe that $p_0 = p$, p_s is decreasing in s, and $p_s \downarrow 0$ as $s \to \infty$. Now the $(s + 1)$th dive will either be successful, in which case the new *a posteriori* probability the treasure is there is one, or it will be unsuccessful, in which case the new *a posteriori* probability p_{s+1} the treasure is there will be less than its previous value p_s. Moreover, if the treasure is (resp., is not) there, in which case all dives are unsuccessful with probability zero (resp., one), then the *a posteriori* probability the treasure is (resp., is not) there converges to one as the number of dives increases without limit. Incidentally, this problem is a prototype of the Bayesian approach to sequential statistical decision problems.

If we (somewhat arbitrarily) limit the number of dives to at most $S + 1$, the above problem becomes one of optimally stopping an $S + 1$ state Markov chain having a stopped state. The states are simply the possible numbers $0, 1, \cdots, S$ of unsuccessful dives. Moreover, for each $s, t = 0, 1, \cdots, S$, one has $r_s = 0$, $c_s = c - rp_s$, $p_{s,s+1} = qp_s + 1 - p_s$ and $q_s = 1 - p_{s,s+1} = (1 - q)p_s$ for $s < S$, $p_{st} = 0$ for $s = S$ or $t \neq s + 1 \leq S$, and $q_S = 1$. Since p_s is decreasing in s, c_s is increasing in s. Thus there is a unique integer s^*, $0 \leq s^* \leq S + 1$, such that $c_s < 0$ for $0 \leq s < s^*$ and $c_s \geq 0$ for $s^* \leq s \leq S$. Evidently, conditions (i) and (ii) above are fulfilled so the myopic policy is optimal. Thus, after s unsuccessful dives, it is optimal to dive again if $s < s^*$ and to quit diving if $s^* \leq s$.

Example 3. Inventory Control.

An inventory manager facing uncertain, independent, non-negative integer demands for a single product over a sequence of periods seeks an ordering policy that will minimize the expected costs incurred before the product becomes obsolete. The probability that the product does not become obsolete in a period given that it was not previously obsolete is β, $0 \leq \beta < 1$. If the product is not obsolete in a period, the probability that the demand is i ($= 0, 1, \cdots$) is p_i where $\Sigma_{i=0}^{\infty} p_i = 1$. The manager can never store more than S units of the product. At the beginning of each period prior to obsolescence, the inventory manager observes the initial stock s ($0 \leq s \leq S$) on hand, the state of the system. There is also a stopped state corresponding to obsolescence. He then places an order that is delivered instantaneously bringing his starting stock in the period up to $a \in A_s \equiv \{s, s + 1, \cdots, S\}$. Demands in the period which cannot be satisfied from the starting stock in the period are simply lost. On taking account of the possibility of obsolescence, the conditional probability $p(t \mid s, a)$ the initial stock in the subsequent period is t equals βp_{a-t} for $0 < t \leq a$, $\beta \Sigma_{i=a}^{\infty} p_i$ for $t = 0$, and zero otherwise. Also $q(s, a) = 1 - \beta$.

There is a cost $K(b)$ of ordering $b \geq 0$ units of stock and a holding and shortage cost $h(a - d)$ in a period when the starting stock is a and demand is d in the period. Then $H(a) = \Sigma_{i=0}^{\infty} h(a - i) p_i$ is the expected holding and shortage cost in the period which we take to exist and be finite. Thus $r(s, a) = -K(a - s) - H(a)$ for $s \leq a$, completing the formulation of the problem as a Markov decision chain.

A good deal can be said about the optimal ordering policy in the important case where $K(0) = 0$ and $K(b) = K \geq 0$ for $b > 0$; and where there is an integer S^*, $0 \leq S^* \leq S$, such that $H(a)$ is nonincreasing in a for $0 \leq a \leq S^*$ and nondecreasing in a for $S^* \leq a \leq S$ (e.g., this is so if $h(\cdot)$, and so $H(\cdot)$, is convex). If $K = 0$, the problem involves balancing storage and shortage costs as discussed in the introduction. In this situation one myopic policy involves choosing $a = \max(S^*, s)$, i.e., ordering up to S^* in every period. Moreover, the myopic policy is optimal because no matter what the (nonnegative) demands in each period, that policy will simultaneously incur as low a value of $H(\cdot)$ in *every* period as can be obtained by any other policy.

If the set-up cost K for placing an order is positive, there is a new dimension to the problem. It is then necessary to strike a balance between the high procurement costs resulting from frequent ordering (e.g., as with the policy in the preceding paragraph) and the high storage and shortage costs ensuing from the greater fluctuations in starting stock levels that accompany infrequent ordering. It turns out that one optimal way to do this is to employ an (s', S') policy, i.e., choose $a = S'$ if $s < s'$ and $a = s$ otherwise, where $0 \leq s' \leq S' \leq S$ are suitably chosen integer constants. The constants s' and S' may be computed in various ways, including, for example, specializing the policy improvement method of Section 3.

One myopic policy for this problem is also of the (s', S') form where $S' = S^*$ and $s' = s^*$ is the smallest nonnegative integer for which $H(s^*) \leqq K + H(S^*)$. Although the myopic policy (s^*, S^*) is not generally optimal, it is related to one optimal (s', S') policy by the inequalities $s^* \leqq s' \leqq S^* \leqq S'$. Thus the optimal policy avoids ordering too frequently by imposing a minimum order size $S' - s'$ while ensuring the negative effects of the fluctuations in starting stocks are minimized by centering them around the desired level S^*.

3. MAXIMUM EXPECTED REWARD

As the examples in the preceding section attest, a basic problem in the study of Markov decision chains is to find a policy that maximizes the expected reward earned before entering the stopped state(s). In this section we formulate this problem precisely and prove the existence of a stationary optimal policy constructively under appropriate conditions.

Let $\Delta = \times_{s=1}^{S} A_s$ be the set of all *decision rules*, i.e., of all functions δ mapping each state s into an action $\delta^s \in A_s$. A *policy* is a sequence $\pi = (\delta_1, \delta_2, \cdots)$ of decision rules, often written for brevity $\pi = (\delta_N)$.[*] Using π means that δ_1 is used at time one, δ_2 is used at time two, and so on. Thus if the system is observed in state s at time N, the action chosen at that time is δ_N^s, the sth component of δ_N. Sometimes, in order to avoid double subcripts we shall use $\delta(N)$ interchangeably with δ_N without mentioning that fact. We write δ^∞ for the *stationary* policy $(\delta, \delta, \cdots)$.

For each decision rule δ, let r_δ be the S element column vector of one period rewards earned by δ. Thus, the sth component of r_δ is $r(s, \delta^s)$. Similarly, let P_δ be the $S \times S$ matrix of one step transition probabilities when δ is used. The stth component of P_δ is $p(t \mid s, \delta^s)$. If $\pi = (\delta_N)$ is a policy, let $P_\pi^N = P_{\delta(1)} \cdots P_{\delta(N)}$ be

---

[*] The reader will note that for simplicity we are restricting attention to nonrandomized policies in which the action chosen in each state at each point in time depends on the time point and the state of the system at that time, but not otherwise upon the prior "history" of the system. It is known that for the optimality criteria we consider, nothing is lost by so doing.

the N-step transition matrix resulting from the use of π. The stth element of P_π^N is the conditional probability that the system is in state t at time $N + 1$ given that it is in state s at time one, when π is used. In particular, $P_\pi^0 = I$ and, if $\pi = \delta^\infty$, $P_\pi^N = (P_\delta)^N \equiv P_\delta^N$, the Nth power of P_δ.

Neumann series.

At this point we need to digress briefly to record a few facts about sequences of matrices. If B_0, B_1, $\cdots$ and B are $S \times S$ real matrices, we say B_N *converges* to B, written $B_N \to B$, if B_N converges elementwise to B, i.e., if the stth element of B_N converges to the stth element of B for all s,t. Similarly, we say the series $\Sigma_{N=0}^\infty B_N$ *converges* if it does so elementwise.

If B is a real $S \times S$ matrix, then a routine computation shows

$$(1) \qquad I - B^N = (I - B)\left(\sum_{i=0}^{N-1} B^i\right), \qquad N = 1, 2, \cdots.$$

Now if $B^N \to 0$, the left hand side of (1) is nonsingular for large enough N. Hence $I - B$ is nonsingular and on premultiplying (1) by $(I - B)^{-1}$ and letting $N \to \infty$ we get the Neumann series expansion

$$(2) \qquad (I - B)^{-1} = \sum_{N=0}^\infty B^N.$$

In order to prepare the way for a needed converse to this result, we give some definitions. If C is a real matrix with nonnegative elements, we say C is *nonnegative* and write $C \geqq 0$. If C and D are real matrices of comparable size, we write $C \leqq D$ if $D - C \geqq 0$. Also write $C < D$ if $C \leqq D$ and $C \neq D$. If C is a real $S \times S$ nonnegative matrix and $C1 \leqq 1$, where 1 is an S element column vector of $+1$'s, we say C is *substochastic*. If further $C1 = 1$, C is called *stochastic*.

If B is substochastic and $I - B$ is nonsingular, we claim $B^N \to 0$, and so (2) holds. To see this, premultiply (1) by $(I - B)^{-1}$ to get

$$(3) \qquad (I - B)^{-1}(I - B^N) = \sum_{i=0}^{N-1} B^i.$$

Since B is nonnegative, the right hand side of (3) is nondecreasing in N. Moreover, because B is substochastic, B^N is substochastic and so the left hand side of (3) is bounded. Hence, the right hand side of (3) converges and so $B^N \to 0$ as desired.

Transient policies.

A policy π is called *transient* if $\Sigma_{N=0}^{\infty} P_\pi^N$ converges. The probabilistic interpretation of this condition is that under π the conditional expected system sojourn time in each state t before entering the stopped state(s) given the system was initially in any state s is finite. To see this, observe that the stth element of the above series is the sum over N of the conditional probability that the system is in state t at time $N + 1$ given it was initially in state s. And this sum is evidently the claimed expected sojourn time.

Since P_δ is substochastic for reach decision rule δ, it follows from the preceding subsection that the stationary policy δ^∞ is transient if and only if $I - P_\delta$ is nonsingular. And in that event we see from (2) that

$$(4) \qquad\qquad (I - P_\delta)^{-1} = \sum_{N=0}^{\infty} P_\delta^N.$$

Thus the stth element of $(I - P_\delta)^{-1}$ is simply the (nonnegative) conditional mean sojourn time in state t under δ^∞ given the system starts in state s.

If $\pi = (\delta_N)$ is transient, then the series

$$(5) \qquad\qquad V_\pi \equiv \sum_{N=0}^{\infty} P_\pi^N r_{\delta(N+1)}$$

converges absolutely because r_δ is bounded in δ. Observe that the sth element $V_{\pi s}$ of the S element column vector V_π is simply the conditional expected reward earned by π before entering the stopped state(s) given that the system starts in state s. To see this, note first that the above conditional expected reward is the sum over N of the corresponding conditional expected rewards earned at each time $N + 1$. But this last conditional expected reward is simply the sth element of $P_\pi^N r_{\delta(N+1)}$, i.e., the sum over t of the conditional probability of entering state t at time $N + 1$ under π starting from s times the reward $r(t, \delta_{N+1}^t)$ earned in state t at time $N + 1$ under π. If $\pi = \delta^\infty$ is transient, it follows from (4) and (5) on putting $V_\delta \equiv V_\pi$ that

$$(6) \qquad V_\delta = \left(\sum_{N=0}^{\infty} P_\delta^N\right) r_\delta = (I - P_\delta)^{-1} r_\delta.$$

In view of (5), a necessary and sufficient condition that the *expected reward* V_π be defined and finite for all policies π and every reward function $r(\cdot, \cdot)$ is that every policy be transient. For this reason we shall focus on this situation, called the *transient case*, in the remainder of this section.

In the transient case, a policy σ is called *optimal* if $V_\sigma \geqq V_\pi$ for all π, i.e., if σ *simultaneously* maximizes the expected reward over all policies starting from *each* possible initial state. Our aim is to establish constructively the existence of a stationary optimal policy. That an optimal policy might be stationary is, perhaps, not too surprising in view of the stationarity of the model. Also, for each initial state s, there is certainly a stationary policy, depending on s, that maximizes $V_{\delta s}$ over δ in the finite set Δ. What is more surprising, however, is that these S maximization problems—one for each initial state s—have a *common* stationary optimal policy.

Comparison Lemma.

The key to our construction is the following fundamental comparison lemma. If γ is a decision rule and $\sigma = (\delta_N)$ is a policy, write $\gamma\sigma$ for the policy $(\gamma, \delta_1, \delta_2, \cdots)$.

LEMMA 1: *If $\pi = (\gamma_N)$ and $\sigma = (\delta_N)$ are transient policies, then*

$$(7) \qquad V_\pi - V_\sigma = \sum_{N=0}^{\infty} P_\pi^N C_{\gamma(N+1)\sigma}$$

where $C_{\gamma\sigma} \equiv V_{\gamma\sigma} - V_\sigma = r_\gamma + P_\gamma V_\sigma - V_\sigma$ for all $\gamma \in \Delta$. If also $\pi = \gamma^\infty$, then

$$(8) \qquad V_\gamma - V_{,\sigma} = (I - P_\gamma)^{-1} C_{\gamma\sigma}.$$

Proof: Let $\pi(N) = (\gamma_1, \cdots, \gamma_N, \delta_1, \delta_2, \cdots)$ and $\pi(0) = \sigma$. Now since σ is transient, $V_{\pi(N)} = \Sigma_{i=0}^{N-1} P_\pi^i r_{\gamma(i+1)} + P_\pi^N V_\sigma$, so because π is transient, $\lim_{N \to \infty} V_{\pi(N)} = V_\pi$. Thus

$$V_\pi - V_\sigma = \lim_{N \to \infty}(V_{\pi(N+1)} - V_{\pi(0)}) = \sum_{N=0}^\infty (V_{\pi(N+1)} - V_{\pi(N)})$$

$$= \sum_{N=0}^\infty P_\pi^N C_{\gamma(N+1)\sigma},$$

which proves (7). The final assertion (8) is then immediate from (4) and (7), completing the proof.

Policy improvement.

We now apply the comparison lemma in the case where each stationary policy is transient. In that event, put $C_{\gamma\delta} \equiv C_{\gamma\pi}$ where $\pi \equiv \delta^\infty$ and $\gamma, \delta \in \Delta$. Since $(I - P_\gamma)^{-1}$ is nonnegative by (4), we have from (8) that $V_\gamma - V_\delta > 0$ (resp., ≤ 0) if the comparison function $C_{\gamma\delta} > 0$ (resp., ≤ 0). For this reason, we say γ is an *improvement* of δ if $C_{\gamma\delta} > 0$.

If there is no improvement of δ, we claim $C_{\gamma\delta} \leq 0$ for all γ, and so $V_\gamma \leq V_\delta$ for all γ. This assertion is an easy consequence of the fact that $C_{\delta\delta} = 0$ and the tth element $C_{\gamma\delta t}$ of $C_{\gamma\delta}$ depends on γ^t but not on γ^s for $s \neq t$. To see this, suppose the claim is false, i.e., $C_{\gamma\delta t} > 0$ for some γ and some state t. Define the decision rule β by $\beta^t = \gamma^t$ and $\beta^s = \delta^s$ for $s \neq t$. Then $C_{\beta\delta t} = C_{\gamma\delta t} > 0$ and $C_{\beta\delta s} = C_{\delta\delta s} = 0$ for $s \neq t$. Hence, $C_{\beta\delta} > 0$ and β is an improvement of δ, which is a contradiction and so establishes our claim.

The above remarks permit us to prove the following result constructively.

LEMMA 2: *If every stationary policy is transient, there is a stationary policy maximizing the expected reward over the class of stationary policies.*

Proof: Let $\delta_0 \in \Delta$ be arbitrary and define the decision rules δ_1, δ_2, $\cdots$ inductively as follows. Given δ_N, let δ_{N+1} be an improvement of δ_N, if one exists, and terminate with δ_N otherwise. Since each successive policy has a higher expected reward than its predecessor, no decision rule can appear twice in the sequence.

Thus because there are only finitely many decision rules, the sequence must terminate with a $\delta \equiv \delta_N$ having no improvement. Hence, from the discussion preceding the Lemma, $C_{\gamma\delta} \leqq 0$ for all γ so δ^∞ maximizes the expected reward over all stationary policies, completing the proof.

The finite algorithm given in the above proof for finding a stationary policy having maximum expected reward is called the *policy improvement method*. The process of finding an improvement of δ, if one exists, involves two steps. First, solve the system $(I - P_\delta)V = r_\delta$ for $V = V_\delta$. Second, find, for each state s, an action $a = \gamma^s \in A_s$ for which $r(s, a) + \Sigma_t\, p(t \mid s, a)V_{\delta t} \geqq V_{\delta s}$ with strict inequality holding for at least one s.

Existence and characterization.

Let $q_\pi^N = (I - P_\pi^N)1$ be the vector of probabilities that the chain enters the stopped state(s) in N steps or less from the various states when π is used. Since P_δ is substochastic for each δ, $P_\pi^N 1$ is nonincreasing in N so $q_\pi^N \equiv (q_{\pi s}^N)$ is nondecreasing in N.

THEOREM 1 (Characterization of transient case): *The following are equivalent*:

   1°   *Every stationary policy is transient.*
   2°   *Every policy is transient.*
   3° q_π^S *is positive for every policy* π.

Proof: $1° \Rightarrow 3°$. Let T_N be the set of states s from which the stopped state(s) are accessible in N steps or less under every policy, i.e., s is stopped or $\min_\pi q_{\pi s}^N > 0$. Thus T_0 is the set of stopped state(s). Let U_N be the set of states $s \notin T_N$ that are accessible in one step to some state in T_N with every action, i.e., for each $a \in A_s$ there is a $t \in T_N$ with $p(t \mid s, a) > 0$. Clearly, $T_{N+1} = T_N \cup U_N$ so $T_0 \subseteq T_1 \subseteq T_2 \subseteq \cdots$. Since there are S states and the U_N are disjoint, at most S of the U_N can be nonempty. Thus $U_N = \phi$ for some $N \leqq S$ and so $T_N = T_{N+1} = \cdots$. If T_N contains every state, then $3°$ holds. In the contrary event, for each state s in the complement of T_N, there is an action $a = a_s \in A_s$ for which $p(t \mid s, a) = = 0$ for all $t \in T_N$. Define $\delta \in \Delta$ by $\delta^s = a_s$ for $s \notin T_N$ and let

δ^s ($\in A_s$) be arbitrary for $s \in T_N$. By construction the stopped state(s) are not accessible from the states in the complement of T_N under δ^∞ so that policy is not transient. This contradicts 1° and so establishes 3°.

3° $\Rightarrow$ 2°. Let π be a policy. By hypothesis, there is a scalar λ, $0 < \lambda < 1$, with $P_\pi^S 1 \leqq \lambda 1$. Thus $P_\pi^{nS} 1 \leqq \lambda^n 1$, so because $P_\pi^N 1$ is nonincreasing in N, $\Sigma_{N=0}^\infty P_\pi^N 1 \leqq S \Sigma_{n=0}^\infty P_\pi^{nS} 1 \leqq S(1 - \lambda)^{-1} 1$. Hence, π is transient as desired.

2° $\Rightarrow$ 1°. This is trivially so, completing the proof.

Remark 1. Notice the proof that 1° implies 3° of Theorem 1 is actually a constructive procedure for determining whether or not every policy is transient, and if not, exhibiting a stationary policy that is not transient.

Remark 2. Observe that 3°, and so 1° and 2° of Theorem 1, are *combinatorial properties* of Markov decision chains, i.e., they depend only on the signs (zero or positive) of the one step transition probabilities.

THEOREM 2 (Existence and characterization): *If every stationary policy is transient, there is a stationary optimal policy. Moreover, a policy σ is optimal if and only if $C_{\gamma\sigma} \leqq 0$ for all $\gamma \in \Delta$.*

Proof: By Lemma 2, there is a stationary transient policy δ^∞ with $C_{\gamma\delta} \leqq 0$ for all γ. Let $\pi = (\gamma_N)$ be any policy. By Theorem 1, π is transient. Thus since $C_{\gamma(N+1)\delta} \leqq 0$ for all N, we have from (7) of the comparison lemma that $V_\pi \leqq V_\delta$, so δ^∞ is optimal. Now if σ is optimal, $V_\sigma = V_\delta$ so $C_{\gamma\sigma} = C_{\gamma\delta} \leqq 0$ for all γ. Conversely, if $C_{\gamma\sigma} \leqq 0$ for all γ, then σ is transient by Theorem 1 and optimal by (7) of the comparison lemma, completing the proof.

Applications.

As an illustration of the techniques of this section, we explore briefly when each of the examples of Section 2 is transient. The gambling problem of example 1 is transient because, in the notation of the proof of Theorem 1, $T_N = \{0, 1, \cdots, N, S - N, S - N + 1, \cdots, S\}$ for $N \leqq S/2$ and $T_N = \{0, 1, \cdots, S\}$ for $N \geqq S/2$. In order for the stopping problem of example 2 to be transient, it is certainly necessary that the stationary policy δ^∞ that continues in each state be transient. This condition is also sufficient. For let α be any other decision rule. Then since $P_\delta = (p_{st})$, $P_\alpha \leq P_\delta$ and so $P_\alpha^N \leq P_\delta^N \to 0$, whence every stationary policy is transient. Finally, the inventory problem of example 3 is transient because $T_1 \supseteq \{0, \cdots, S\}$.

4. DISCOUNT OPTIMALITY

In many applications of Markov decision chains it is unsatisfactory to compare policies on the basis of their expected rewards because they are infinite or even undefined. For example, that is usually the case for the inventory problem discussed in Example 3 of Section 2 when there is no risk of obsolescence. In these circumstances one is led to consider more suitable optimality criteria.

Maximum expected discounted reward.

One mathematically satisfactory and economically meaningful approach is to discount future rewards. The idea is to assume the (possibly negative) net income received in each period is deposited in or withdrawn from a bank whose interest rate on both savings and loans is $100\rho\%$ $(\rho > 0)$ per period. At the given interest rate, a deposit of $\beta = (1 + \rho)^{-1}$ dollars in one period will grow, with interest, to one dollar in the following period. Similarly, a deposit of β^N dollars in one period will grow to one dollar N periods later. Thus, the size V_π^ρ of the deposit needed at time zero to produce the (bounded) sequence of expected rewards $\{P_\pi^{N-1} r_{\delta(N)}\}_{N=1}^\infty$ at times $1, 2, \cdots$ under the policy $\pi = (\delta_N)$ is

$$(9) \qquad V_\pi^\rho \equiv \sum_{N=1}^\infty \beta^N P_\pi^{N-1} r_{\delta(N)}.$$

Notice that since each P_π^N is substochastic and r_δ is bounded in δ, the series converges absolutely for all π. We call V_π^ρ the *expected discounted reward* from π and seek a policy maximizing V_π^ρ.

This problem is easily reduced to that of Section 3 by letting $\tilde{P}_\delta \equiv \beta P_\delta$ and $\tilde{r}_\delta \equiv \beta r_\delta$ for all $\delta \in \Delta$. In effect this replaces the discounted problem by an undiscounted one in which, like the inventory problem of Section 2, there is an obsolescence probability $1 - \beta$ at each time. Define $\tilde{P}_\pi^N$ as usual except $\tilde{P}_\delta$ replaces P_δ for each δ. Then (9) can be rewritten as

$$(10) \qquad \tilde{V}_\pi = \sum_{N=0}^\infty \tilde{P}_\pi^N \tilde{r}_{\delta(N+1)}.$$

where $\tilde{V}_\pi \equiv V_\pi^\rho$. Since P_δ is substochastic for each δ and $0 \leqq \beta < 1$, it follows that each policy is transient (with the $\tilde{P}_\delta$), and so all results of Section 3 apply at once. In particular, there is a stationary policy maximizing $\tilde{V}_\pi$ over all policies, and one such policy may be found by policy improvement.

Discount Optimality.

If one is uncertain about the exact value of a small interest rate, or if one is really interested in the undiscounted problem, or if the interest rate is so small that the policy improvement method is numerically unstable, it might be of interest to find a *discount optimal* policy σ. By this we mean that for some $\rho^* > 0$,

$$(11) \qquad V_\sigma^\rho \geqq V_\pi^\rho \text{ for all policies } \pi \text{ and all } 0 < \rho < \rho^*,$$

i.e., σ simultaneously maximizes the expected discounted reward for all positive interest rates below ρ^*.

In the remainder of this section we prove constructively the existence of a stationary policy $\sigma = \delta^\infty$ that satisfies (11) for all stationary $\pi = \gamma^\infty$. Then, by the remarks of the preceding subsection, (11) holds for all policies π. For this reason it will suffice to consider only stationary policies in the sequel.

Our approach is to generalize the policy improvement method. Instead of seeking a decision rule that is an improvement for a single fixed interest rate, we seek one that is an improvement for all small enough positive interest rates. To carry this out, we shall need the Laurent expansion of $V_\delta^\rho \equiv V_\pi^\rho$ in ρ about the origin for each $\delta \in \Delta$ where $\pi = \delta^\infty$.

To this end, observe from (6) that on putting $Q_\delta \equiv P_\delta - I$ and

$$R_\delta^\rho = (\rho I - Q_\delta)^{-1},$$

$$(12) \qquad V_\delta^\rho = \left(\sum_{N=0}^\infty \beta^{N+1} P_\delta^N \right) r_\delta = \beta(I - \beta P_\delta)^{-1} r_\delta = R_\delta^\rho r_\delta$$

for all $\rho > 0$. The matrix function R_δ^ρ, called the *resolvent* of Q_δ, is well defined for $\rho > 0$ because $I - \beta P_\delta$, and so $\rho I - Q_\delta$, is non-singular for $\rho > 0$ by (4).

It is clear from (12) that in order to obtain a Laurent expansion of V_δ^ρ, we need one for R_δ^ρ. This could be achieved with the aid of the theory of complex variables, but we shall instead follow a simpler approach that exploits the fact P_δ is substochastic. To achieve this, we need some preliminary results.

Neumann series: (C, 1) limits.

If $B_0, B_1, \cdots$ and B are $S \times S$ real matrices and $B_N' \equiv \Sigma_{i=0}^N B_i$, we write $B_N \to B$ $(C, 1)$ if $N^{-1} B_{N-1}' \to B$. Incidentally, $(C, 1)$ stands for 'Cesaro limit of order one.' Similarly, we write $\Sigma_{N=0}^\infty B_N = B$ $(C, 1)$ if $B_N' \to B$ $(C, 1)$, or equivalently, $N^{-1} \Sigma_{i=0}^{N-1} B_i' \to B$.

On averaging (1) and putting $B_N' \equiv \Sigma_{i=0}^N B^i$ we get

$$(13) \quad I - N^{-1} B_{N-1}' B = (I - B)\left(N^{-1} \sum_{j=0}^{N-1} B_j' \right), \quad N = 1, 2, \cdots.$$

Thus if $B^N \to 0$ $(C, 1)$, we see that for large enough N, the left hand side of (13) is nearly I and so is nonsingular. Thus $I - B$ is nonsingular so on premultiplying (13) by $(I - B)^{-1}$ and letting $N \to \infty$ we get the $(C, 1)$ version of the Neumann series expansion

$$(14) \qquad (I - B)^{-1} = \sum_{N=0}^\infty B^N \ (C, 1).$$

In order to appreciate the significance of this result, observe that since $B^N \to 0$ implies $B^N \to 0$ $(C, 1)$, it follows that (14) will certainly hold whenever $B^N \to 0$. But (14) holds in other cases as well. For example, if $B = -I$, then $B^N \to 0$ $(C, 1)$ even though $B^N \not\to 0$. In that event (14) asserts $(\frac{1}{2})I = \Sigma_{N=0}^\infty (-I)^N$ $(C, 1)$ though, of course, the last series does not converge in the usual sense.

Stationary and deviation matrices.

As an application of (14), suppose P is an $S \times S$ substochastic matrix. We show first that there is a unique substochastic matrix P^*, called the *stationary matrix*, such that

(15) $$P^N \to P^* \; (C, 1),$$

and moreover,

(16) $$PP^* = P^*P = P^*P^* = P^*.$$

To see this consider the average $B_N \equiv N^{-1} \sum_{j=0}^{N-1} P^j$. Since P is substochastic, so is P^N and B_N. Thus $\{P^N\}$ and $\{B_N\}$ are bounded. Hence, to prove (15), it will suffice to show that the limit points of $\{B_N\}$ coincide. (A limit point of a sequence of matrices is, of course, the limit of a subsequence of the matrices.) Let J be any limit point of $\{B_N\}$. Then $B_N + N^{-1}(P^N - I) = PB_N = B_N P$ have the common limit point $J = PJ = JP$. Thus $J = P^N J = JP^N$ and so $J = B_N J = JB_N$. Now let K be any limit point of $\{B_N\}$. Then the preceding equalities have the common limit point $J = KJ = JK$. Since J and K are arbitrary limit points, they can be interchanged giving $K = JK = KJ$. Thus $J = K$ so (15) holds. Moreover, in the course of the above development we have also established (16), completing the proof.

In view of (15), the stth element of P^* can be interpreted as the long run expected fraction of the time the associated Markov chain spends in state t given that it started in state s. Notice from $P^*P = P^*$ in (16) that each row of P^* is left stationary when postmultiplied by P. Thus, by (15), the sth row of P^* is the stationary distribution of the associated Markov chain to which the N-step transition probabilities from state s converge $(C, 1)$.

It follows readily from (16) that

(17) $$(P - P^*)^N = P^N - P^* \quad \text{for} \quad N \geqq 1.$$

On setting $B = P - P^*$, it is apparent from (15) and (17) that $B^N \to 0 \; (C, 1)$. Hence on putting $Q \equiv P - I$, we see by (14) that

$P^* - Q = I - B$ is nonsingular and, by (14) and (17),

$$(P^* - Q)^{-1} = \sum_{N=0}^{\infty} (P - P^*)^N = \sum_{N=0}^{\infty} (P^N - P^*) + P^* \ (C, 1).$$

On premultiplying this equation by $(I - P^*)$, using (16), and putting $D \equiv (I - P^*)(P^* - Q)^{-1}$, we get

$$(18) \qquad\qquad D = \sum_{N=0}^{\infty} (P^N - P^*) \ (C, 1).$$

Notice from (16) and (18) that the stth element of D is the $(C, 1)$ limit of the amount by which the expected N period sojourn time of the associated Markov chain in state t when the system starts in state s exceeds that when the system starts instead with the stationary distribution for state s as the initial state probability vector. This interpretation suggests we call D the *deviation matrix*.

Laurent Expansion of Resolvent.

Let $R^\rho \equiv (\rho I - Q)^{-1}$ be the resolvent of Q for $\rho > 0$. We are now ready to develop R^ρ in a Laurent series about the origin whose coefficients are the stationary matrix P^* and powers of the deviation matrix D.

THEOREM 3: *For all small enough $\rho > 0$,*

$$(19) \qquad\qquad R^\rho = \rho^{-1}P^* + \sum_{n=0}^{\infty} (-\rho)^n D^{n+1}.$$

Proof: It follows from (16) that $QP^* = P^*Q = 0$. This implies that

$$(20) \qquad\qquad R^\rho P^* = P^* R^\rho = \rho^{-1}P^*$$

and that $I - P^*$ and $(P^* - Q)^{-1}$ commute. The last fact and (16) imply that

$$(21) \qquad\qquad DP^* = P^*D = 0$$

and

$$(22) \quad (I - P^*)(I + \rho D) = (P^* - Q)D + \rho D = (\rho I - Q)D.$$

Now for all small enough $\rho > 0$, $I + \rho D$ is nonsingular, and by (2) its inverse has a Neumann series expansion. Thus on premultiplying (22) by R^ρ and postmultiplying (22) by $(I + \rho D)^{-1}$ we get

$$(23) \qquad R^\rho(I - P^*) = D(I + \rho D)^{-1} = \sum_{n=0}^{\infty} (-\rho)^n D^{n+1}.$$

Adding (20) and (23) completes the proof.

Characterization of Discount Optimal Policies.

It follows from (12) and Theorem 3 that for all small enough $\rho > 0$,

$$(24) \qquad V_\delta^\rho = \sum_{n=-1}^{\infty} \rho^n v_\delta^n$$

where $v_\delta^{-1} \equiv P_\delta^* r_\delta$ and $v_\delta^n \equiv (-1)^n D_\delta^{n+1} r_\delta$ for $n = 0, 1, \cdots$, with P_δ^* and D_δ being respectively the stationary and deviation matrices determined by P_δ. Evidently (24) gives the desired Laurent expansion of V_δ^ρ in ρ.

Before attempting to compare policies, we need a definition. A real matrix B is called *lexicographically nonnegative*, written $B \geq 0$, if the first nonvanishing element (if any) of each row of B is positive. Similarly write $B > 0$ if $B \geq 0$ and $B \neq 0$. Also write $B \leq 0$ (resp., $B < 0$) if $-B \geq 0$ (resp., $-B > 0$).

Let $V_\delta \equiv (v_\delta^{-1}, v_\delta^0, v_\delta^1, \cdots)$. Then from (24),

$$(25) \qquad V_\delta^\rho - V_\gamma^\rho = \sum_{n=-1}^{\infty} \rho^n (v_\delta^n - v_\gamma^n)$$

is nonnegative for all small enough $\rho > 0$ if and only if $V_\delta - V_\gamma \geq 0$. The reason for this is that for small enough positive ρ, the sign of each row of (25) is simply the sign of the first nonvanishing coefficient of the Laurent expansion. It follows from this fact that δ^∞ is discount optimal if and only if $V_\delta - V_\gamma \geq 0$ for all $\gamma \in \Delta$.

Sensitive policy improvement.

We are now in a position to generalize the policy improvement method to find a discount optimal policy. It follows by application of the comparison lemma to the discounted problem that for $\rho > 0$

$$(26) \qquad V_\gamma^\rho - V_\delta^\rho = R_\gamma^\rho C_{\gamma\delta}^\rho,$$

where $C_{\gamma\delta}^\rho \equiv r_\gamma + Q_\gamma V_\delta^\rho - \rho V_\delta^\rho$. Also on substituting the formula (24) for V_δ^ρ in the definition of the *comparison function* $C_{\gamma\delta}^\rho$ we get

$$(27) \qquad C_{\gamma\delta}^\rho = \sum_{n=-1}^{\infty} \rho^n c_{\gamma\delta}^n$$

for all small enough $\rho > 0$ where $c_{\gamma\delta}^n \equiv r_\gamma^n + Q_\gamma v_\delta^n - v_\delta^{n-1}$ for $n = -1, 0, 1, \cdots$, $r_\gamma^n = 0$ for $n \neq 0$, $r_\gamma^0 = r_\gamma$, and $v_\delta^{-2} = 0$. Now put $C_{\gamma\delta} = (c_{\gamma\delta}^{-1}, c_{\gamma\delta}^0, c_{\gamma\delta}^1, \cdots)$. Observe that $C_{\gamma\delta}^\rho > 0$ (resp., $\leqq 0$) for all small enough $\rho > 0$ if and only if $C_{\gamma\delta} > 0$ (resp., $C_{\gamma\delta} \leq 0$). Since $R_\delta^\rho > 0$ for $\rho > 0$, it follows from (25)–(27) that $V_\gamma - V_\delta > 0$ (resp., ≤ 0) if $C_{\gamma\delta} > 0$ (resp., ≤ 0). For this reason we say γ is a *sensitive improvement* of δ if $C_{\gamma\delta} > 0$. If there is no sensitive improvement of δ, then because $C_{\delta\delta}^\rho = 0$ for all $\rho > 0$ by (26), and so $C_{\delta\delta} = 0$ by (27), we see just as in Section 3 that $C_{\gamma\delta} \leq 0$ for all γ and so $V_\delta - V_\gamma \geq 0$ for all γ. Using these facts, we can now prove the main result of this section.

THEOREM 4 (Existence and characterization): *There is a stationary discount optimal policy. Also the stationary policy δ^∞ is discount optimal if and only if $C_{\gamma\delta} \leq 0$ for all γ.*

Proof: Let $\delta_0 \in \Delta$ be arbitrary and choose $\delta_1, \delta_2, \cdots$ in Δ inductively as follows. Given δ_N, let δ_{N+1} be a sensitive improvement of δ_N, if one exists, and terminate with δ_N otherwise. Since $V_{\delta(N)}$ increases lexicographically with N, no decision rule can appear twice in the sequence. Thus because there are only finitely many decision rules, the sequence must terminate with a $\delta \equiv \delta_N$ having no improvement. But then from the discussion preceding the Theorem, $C_{\gamma\delta} \leq 0$ for all γ and $V_\delta - V_\gamma \geq 0$ for all γ so δ^∞ is discount optimal. The final assertion of the theorem follows by noting that if $C_{\gamma\delta} \nleq 0$ for some γ, then $C_{\alpha\delta} > 0$ for some α so δ^∞ is not optimal. This completes the proof.

The finite algorithm given in the proof for finding a stationary discount optimal policy is called the *sensitive policy improvement* method. Each iteration involves two steps. For a given δ, first one computes V_δ. Then one seeks a γ with $C_{\gamma\delta} > 0$.

Truncated matrices.

As described above, each iteration of the sensitive policy improvement method appears to require computation of the infinite matrices V_δ and $C_{\gamma\delta}$. It turns out, however, that it suffices to compute the finite truncated matrices $V_\delta^n \equiv (v_\delta^{-1}, \cdots, v_\delta^n)$ and $C_{\gamma\delta}^n \equiv (c_{\gamma\delta}^{-1}, \cdots, c_{\gamma\delta}^n)$ for $n = S - N$ where N is the rank of P_δ^*, or equivalently, the number of recurrent classes in P_δ.

In order to prove this, we require a preliminary result.

LEMMA 3: *If B is a real $S \times S$ matrix, L is a subspace of R^S, and $B^i x \in L$ for $i = 1, \cdots, d$, where d is the smaller of the rank of B and one plus the dimension of L, then $B^i x \in L$ for $i = 1, 2, \cdots$.*

Proof: We show first that there is a nonnegative integer $j \leq d$ such that $B^{j+1}x$ is a linear combination of $0, B^1 x, \cdots, B^j x$. If d equals one plus the dimension of L, then because the number of linearly independent vectors in a subspace cannot exceed its dimension, the vectors $B^1 x, \cdots, B^d x$ are linearly dependent and so the assertion holds. If instead d equals the rank of B, then because the dimension of the subspace spanned by the columns of B is d, we see that $B^1 x, \cdots, B^{d+1}x$ are linearly dependent, from which the assertion follows.

We now show by induction on k that $B^k x$ is a linear combination of $0, B^1 x, \cdots, B^j x$ for all $k \geq 1$, which, because L is a subspace, will complete the proof. This is so by construction for $1 \leq k \leq j + 1$. Suppose it holds for $k - 1$ ($\geq j + 1$) so $B^{k-1}x = \Sigma_{i=1}^j \lambda_i B^i x$. Premultiplying this equation by B gives $B^k x = \Sigma_{i=1}^j \lambda_i B^{i+1}x$. Since $B^{j+1}x$ is a linear combination of $0, B^1 x, \cdots, B^j x$, so is $B^k x$, completing the proof.

We need one more fact before coming to the promised truncation theorem. Observe from (21) and $P_\delta^* + D_\delta = (P_\delta^* - Q_\delta)^{-1}$ that R^S is the direct sum of the ranges of P_δ^* and D_δ. Thus the sum of the ranks of P_δ^* and D_δ is S.

THEOREM 5: *Suppose γ, $\delta \in \Delta$ and P_δ^* has rank N. Then*

(a) $C_{\gamma\delta} = 0$ *if and only if* $C_{\gamma\delta}^{S-N} = 0$.
(b) $V_\gamma = V_\delta$ *if and only if* $V_\gamma^{S-N} = V_\delta^{S-N}$.

Proof: For part (a), it suffices to show the "if" part. Observe that since $c_{\delta\delta}^n = 0$,

$$(28) \qquad c_{\gamma\delta}^n = c_{\gamma\delta}^n - c_{\delta\delta}^n = (P_\gamma - P_\delta)v_\delta^n, \qquad n = 1, 2, \cdots.$$

Because $C_{\gamma\delta}^{S-N} = 0$, it follows from (28) that

$$(29) \qquad (P_\gamma - P_\delta)v_\delta^n = 0, \qquad n = 1, \cdots, S - N.$$

In view of (28), it suffices to show (29) holds for $n = 1, 2, \cdots$. That this is so follows by an application of Lemma 3 with $B = -D_\delta$, L the null space of $P_\gamma - P_\delta$, and $x = v_\delta^0$, on noting that because the rank of P_δ^* is N, the rank of D_δ is $S - N$.

For part (b), it suffices to show the "if" part. The assertion will be so if

$$(30) \qquad (D_\gamma - D_\delta)v_\delta^n = 0, \qquad n = 0, 1, \cdots.$$

By hypothesis, (30) holds for $n = 0, \cdots, S - N - 1$. That (30) holds for $n = 0, 1, \cdots$ then follows from Lemma 3 with $B = -D_\delta$, L the null space of $D_\gamma - D_\delta$, and $x = -r_\delta$, on noting, as above, that the rank of D_δ is $S - N$. This completes the proof.

Remark. Part (a) of Theorem 5 implies that if the sth, say, row of $C_{\gamma\delta}^{S-N}$ vanishes, the same is so of the sth row of $C_{\gamma\delta}$. As a consequence, if γ is a sensitive improvement of δ, then $C_{\gamma\delta}^{S-N} > 0$. Thus, in searching for a sensitive improvement of δ, it is never necessary to look beyond the matrices V_δ^{S-N} and $C_{\gamma\delta}^{S-N}$ for $\gamma \in \Delta$. Observe also that $N = 0$ if and only if P_δ is transient. Otherwise $N \geqq 1$.

Because the computations required to find a discount optimal policy are formidable, it is often of interest to substitute simpler and less sensitive discount criteria. We explore some of these now, and as a by-product, obtain efficient ways of implementing the sensitive policy improvement method.

5. n DISCOUNT OPTIMALITY

For each $n = -1, 0, 1, \cdots$ a policy σ will be called n *discount optimal* if

(31) $\underline{\lim}_{\rho \downarrow 0} \rho^{-n} (V_\sigma^\rho - V_\pi^\rho) \geqq 0$ for all policies π.

Observe first that if (31) holds for n, then it surely holds for $n - 1$. Thus if σ is n discount optimal, then σ is also $n - 1$ discount optimal. Furthermore, if σ is discount optimal, then σ is certainly n discount optimal for all n. Thus as n increases, the n discount optimality criterion becomes more sensitive.

Since there is a stationary discount optimal policy, that policy is also n discount optimal. Thus we can restrict attention to stationary policies in the sequel without loss of optimality. Let $\mathfrak{D}_n$ (resp., $\mathfrak{D}$) be the set of all decision rules δ for which δ^∞ is n discount (resp., discount) optimal. It follows from (25) and (31) that $\mathfrak{D}_n = \{\delta \in \Delta : V_\delta^n - V_\gamma^n \geq 0 \text{ for all } \gamma \in \Delta\}$ and $\mathfrak{D} = \{\delta \in \Delta : V_\delta - V_\gamma \geq 0 \text{ for all } \gamma \in \Delta\}$. As we have seen, $\mathfrak{D}_{-1} \supseteq \mathfrak{D}_0 \supseteq \cdots \supseteq \mathfrak{D}$. Moreover, by (b) of Theorem 5, $\mathfrak{D}_S = \mathfrak{D}_{S+1} = \cdots = \mathfrak{D}$. Indeed if no stationary policy is transient, we also have $\mathfrak{D}_{S-1} = \mathfrak{D}_S$ because then each P_δ has a recurrent class.

n improvement and characterization of n discount optimality.

Since δ^∞ is n discount optimal if and only if δ maximizes V_δ^n lexicographically, it seems best when seeking such a policy to focus in turn on finding elements of $\mathfrak{D}_{-1}, \mathfrak{D}_0, \cdots, \mathfrak{D}_n$ in that order. We now show how to refine the sensitive policy improvement method to do this.

We say $\gamma \in \Delta$ is an n *improvement* of $\delta \in \Delta$ if $C_{\gamma\delta}^n > 0$ and $C_{\gamma\delta s}^n = 0$ implies $\gamma^s = \delta^s$. Observe that an n improvement of δ is also a sensitive improvement of δ. If there is no n·improvement of δ, then, as usual, $C_{\gamma\delta}^n \leq 0$ for all γ.

THEOREM 6: *If δ has no $n + 1$ improvement, then δ^∞ is n discount optimal. If γ is an n improvement of δ, then $V_\gamma^n - V_\delta^n > 0$.*

Proof: We begin with the first assertion. The hypothesis implies

$C_{\gamma\delta}^{n+1} \leq 0$ for all $\gamma \in \Delta$. Thus from (27), $\varliminf_{\rho\downarrow0} - \rho^{-n-1} C_{\gamma\delta}^{\rho} \geq 0$ for all $\gamma \in \Delta$. Recall $R_{\gamma}^{\rho} > 0$ for all $\rho > 0$, and from Theorem 3, $\varliminf_{\rho\downarrow0} \rho R_{\gamma}^{\rho} = P_{\gamma}^{*}$. Putting these facts together, we get from (26) that

$$\varliminf_{\rho\downarrow0} \rho^{-n}(V_{\delta}^{\rho} - V_{\gamma}^{\rho}) = \varliminf_{\rho\downarrow0} (\rho R_{\gamma}^{\rho})(-\rho^{-n-1}C_{\gamma\delta}^{\rho}) \geq 0$$

for all γ so $\delta \in \mathfrak{D}_n$ as claimed.

For the final assertion, observe that the hypothesis implies that $C_{\gamma\delta}^{n} > 0$ and $C_{\gamma\delta} > 0$. It then follows from (27) that $\varliminf_{\rho\downarrow0} \rho^{-n}C_{\gamma\delta}^{\rho} > 0$ and $C_{\gamma\delta}^{\rho} > 0$ for all small enough $\rho > 0$. Also from (12), $R_{\gamma}^{\rho} \geq \beta I$ for $\rho > 0$. Combining these facts we see from (26) that

$$\rho^{-n}(V_{\gamma}^{\rho} - V_{\delta}^{\rho}) \geq \beta\rho^{-n}C_{\gamma\delta}^{\rho} > 0$$

for all small enough $\rho > 0$. Thus $\varliminf_{\rho\downarrow0} \rho^{-n}(V_{\gamma}^{\rho} - V_{\delta}^{\rho}) \geq \varliminf_{\rho\downarrow0} \rho^{-n}C_{\gamma\delta}^{\rho} > 0$, and so from (25), $V_{\gamma}^{n} - V_{\delta}^{n} > 0$, completing the proof.

An algorithm.

Consider the following algorithm for inductively constructing elements of $\mathfrak{D}_{-1}, \mathfrak{D}_0, \cdots, \mathfrak{D}_n$. Let $\mathfrak{D}_{-2} \equiv \Delta$ and suppose for some $n - 1 \geq -2$ that one has an element δ_{n-1} of $\mathfrak{D}_{n-1}$. (To start the process when $n = -1$, any decision rule δ_{-2} will do because then trivially $\delta_{-2} \in \mathfrak{D}_{-2}$.) Next find an n improvement of δ_{n-1}, or when this is impossible, an $n + 1$ improvement of δ_{n-1}. Then replace δ_{n-1} by its n or $n + 1$ improvement and repeat the procedure. After finitely many repetitions of this process, one must terminate with a δ_n having no $n + 1$ improvement. Consequently, by Theorem 6, $\delta_n \in \mathfrak{D}_n$ as desired. Incidentally, when $n = S - 1$, we can strengthen this conclusion to $\delta_{S-1} \in \mathfrak{D}_S \ (=\mathfrak{D})$ by Theorem 5.

Computations.

Observe that if γ is an n or $n + 1$ improvement of $\delta \equiv \delta_{n-1} \in \mathfrak{D}_{n-1}$, then $C_{\gamma\delta}^{n-1} = 0$. For otherwise γ is an $n - 1$ improvement of δ whence by Theorem 6 we contradict the hypothesis that $\delta \in \mathfrak{D}_{n-1}$. It follows, therefore, that in seeking an n or $n + 1$ improvement of δ, one only needs to compute $c_{\gamma\delta}^{n}$ and, for an $n + 1$ improvement, also $c_{\gamma\delta}^{n+1}$. To do this one must compute v_{δ}^{n}, and when seeking an $n + 1$ improvement, v_{δ}^{n+1} as well. Each of these vectors can be found by a computational procedure requiring an amount of work essentially equal to that of solving a single system of S linear equations in S unknowns. The procedure, which for brevity we omit, avoids computation of P_δ^* or D_δ.

At this point it seems appropriate to review briefly the economic interpretation and significance of n discount optimality. We do this for $n = -1, 0,$ and 1, leaving the other cases to the reader.

−1 discount optimality: earning rate.

Suppose you deposit V_δ^ρ dollars in a bank at time zero and ask that the bank pay you the interest due on this sum at times $1, 2, \cdots$ as a perpetual annuity. Thus ρV_π^ρ can be thought of as the long run average expected reward per period that is equivalent in value to the initial deposit V_π^ρ at time zero. From this viewpoint, -1 discount optimality is equivalent to maximizing the limit of the equivalent average expected reward ρV_δ^ρ per period as the interest rate approaches zero, or briefly the *earning rate*. But from the characterization of $\mathfrak{D}_{-1}$, this is the same as maximizing $v_\delta^{-1} = \lim_{\rho \downarrow 0} \rho V_\delta^\rho$ over all decision rules δ. From (15) and the definition $v_\delta^{-1} = P_\delta^* r_\delta$, maximizing the earning rate is the same as maximizing the long run (time) average expected reward.

If there is a unique stationary -1 discount optimal policy, that policy is necessarily discount optimal as well, and so the -1 discount optimality criterion is eminently satisfactory. But in many— perhaps most—practical problems, there are several -1 discount optimal policies. In that event the criterion has the important limitation that it is indifferent between alternate -1 discount optimal policies which may have major differences in transient earnings. To illustrate, in the example of Figure 1 in the introduction, the earning rate for both policies is the same, viz., two dollars per period, so both are -1 discount optimal. But at each point in time one policy's cumulative earnings exceeds the others by at least ten million dollars starting from state one! The important point about this example is that it is not pathological, except perhaps for the relative magnitude of the transient earnings. It is merely typical of a wide class of practical problems in which there is a stationary -1 discount optimal policy having both transient and recurrent states, the latter all belonging to the same class. In these circumstances, one can arbitrarily alter the actions chosen in the transient states, provided only that those states remain transient, and still be assured that the resulting policy is -1 discount optimal. Yet certain actions in the transient states could produce much larger earnings than others. This situation is illustrated in a practical setting with a -1 discount optimal (s', S') policy in example 3 where there is no risk of obsolescence. Then the inventory levels $S' + 1, \cdots, S$ are transient states so one can use *any* feasible ordering policy in those states (provided there is a positive probability of a demand of size $S - S'$ or larger) and be assured the resulting policy remains -1 discount optimal. These examples suggest the need for an optimality criterion that is sensitive to these transient earnings, e.g., 0 discount optimality.

0 discount optimality: transient earnings.

A policy δ^∞ is 0 discount optimal if it comes arbitrarily close to maximizing the expected discounted reward for all small enough positive interest rates. For the example of Figure 1, the only 0 discount optimal policy is the one which earns ten million dollars in state one.

In the transient case V_δ^ρ converges to the expected reward from δ^∞ as the interest rate converges to zero. In that event -1 discount optimality is the same as maximizing the expected reward.

As we have seen from the characterization of $\mathfrak{D}_0$, a policy δ^∞ is 0 discount optimal if and only if it maximizes $\rho^{-1} v_\delta^{-1} + v_\delta^0$ for all small enough $\rho > 0$. This is equivalent to saying δ maximizes the *transient earnings* $v_\delta^0 = \lim_{\rho \downarrow 0} (V_\delta^\rho - \rho^{-1} v_*^{-1})$ over the set $\mathfrak{D}_{-1}$ of decision rules δ achieving the maximum earning rate $v_*^{-1} = \max_{\delta \in \Delta} v_\delta^{-1}$. It is this secondary maximization that assures the expected transient earnings are as large as possible among those policies having maximum earning rate.

The 0 discount criterion can also be given a time average interpretation. Let $V_\delta^{(N)} = \Sigma_{j=0}^{N-1} P_\delta^{j-1} r_\delta$. Then by (15) and (18), $V_\delta^{(N)} - N v_\delta^{-1} \to v_\delta^0$ $(C, 1)$. Hence δ^∞ is 0 discount optimal if and only if it maximizes the $(C, 1)$ limit of $V_\delta^{(N)} - N v_*^{-1}$ among those δ achieving the maximum long run (time) average expected reward.

The 0 discount optimality criterion places equal weight on earnings in different periods. For example, in the problem of Figure 2 below, both actions α and β in state one are 0 discount optimal, since each earns a total of zero dollars. On the other hand action α, which earns one thousand dollars at time one and loses that sum at time two, seems preferable. The reason is that α's temporary earnings may be used for other purposes, e.g., investment in other activities. This suggests the use of a more refined criterion, like 1 discount optimality, that reflects the value of temporary earnings.

1 discount optimality: temporary earnings.

Suppose you deposit the rewards earned by a Markov decision chain in a bank whose interest rate on savings and on loans is $100p\%$ $(p > 0)$ per period. This bank pays or collects interest when accrued only on your bank balance at each time point *excluding* previous interest earned or paid. Interest paid or collected by the first bank is deposited in or withdrawn from a second bank whose interest rate on both savings and loans is $100\rho\%$ $(\rho > 0)$ per period.

The expected balance at the first bank at time N when δ^∞ is used is the expected N period reward $V_\delta^{(N)}$. The expected discounted value at time -1 of the interest payments deposited in or with-

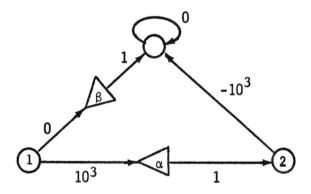

Fig. 2.

drawn from the second bank is $p\bar{V}_\delta^\rho$ where

$$(32) \qquad \bar{V}_\delta^\rho \equiv \sum_{N=1}^\infty \beta^{N+1} V_\delta^{(N)} = \sum_{N=1}^\infty \beta^N V_\delta^\rho = \rho^{-1} V_\delta^\rho.$$

The second equality in (32) is obtained by using (12) to eliminate V_δ^ρ and then collecting terms involving like powers of β. The final equality in (32) is simply the usual formula for the geometric series.

The equations (32) provide the following interpretation of 1 discount optimality. Whatever the interest rate $100p\%$ $(p > 0)$ at the first bank, δ^∞ is 1 discount optimal if and only if $\varliminf_{\rho \downarrow 0}$ $(\bar{V}_\delta^\rho - \bar{V}_\gamma^\rho) \geqq 0$ for all $\gamma \in \Delta$. It follows from this fact and the definitions of $\mathfrak{D}_0$ and $\mathfrak{D}_1$ that 1 discount optimality is concerned with choosing from among all 0 discount optimal policies one giving the largest possible temporary availability of funds (measured in reward-periods) for other purposes.

In particular, suppose in the example of Figure 2 that $\delta^1 = \alpha$ and $\gamma^1 = \beta$. Then $\varliminf_{\rho \downarrow 0}$ $(\bar{V}_{\delta 1}^\rho - \bar{V}_{\gamma 1}^\rho) = 10^3 > 0$ so δ^∞ is 1 discount optimal but γ^∞ is not. The reason is simply that although both policies earn zero total rewards starting from state one, δ^∞ makes one thousand dollars available for one period while γ^∞ does not.

Systems of Markov decision chains.

Another motivation for the study of n discount optimality criteria for $n > -1$ arises from the consideration of an infinite collection of Markov decision chains with staggered starting times. In these circumstances it turns out that in order to assure n discount optimality for the system as a whole, it is often necessary to achieve $n + 1$ discount optimality for each individual chain. In particular to get -1 (resp., 0) discount optimality at the system level, one needs 0 (resp., 1) discount optimality at the chain level. To see this it is necessary to formulate the problem precisely.

Suppose one is interested in controlling a sequence of Markov decision chains labeled $1, 2, \cdots$. For simplicity, assume each chain is chosen by nature from a given finite set of possible chains, and that chain N is started at time N $(= 1, 2, \cdots)$. Let p_N be the row vector of probabilities that chain N starts in each of the possible states at time N. Let $V^{\rho N}_{\pi(N)}$ be the vector of expected discounted rewards earned starting from each state by chain N when the policy $\pi(N)$ is used for that chain, the interest rate is $100\rho\%$ $(\rho > 0)$, and rewards are discounted to time $N - 1$ $(= 0, 1, \cdots)$.

A *strategy* is a sequence $\pi = (\pi(1), \pi(2), \cdots)$ of policies with $\pi(N)$ being the policy used by chain N. Let $\tilde{V}^{\rho}_{\pi} \equiv \Sigma^{\infty}_{N=1} \beta^N p_N V^{\rho N}_{\pi(N)}$ be the expected discounted value at time -1 of the system rewards when the strategy $\pi = (\pi(N))$ is used. A strategy $\sigma = (\sigma(N))$ is called n *discount optimal* for the system if $\lim_{\rho \downarrow 0} \rho^{-n}(\tilde{V}^{\rho}_{\sigma} - \tilde{V}^{\rho}_{\pi}) \geqq 0$ for all strategies π and all initial probabilities (p_N) where $n = -1, 0, 1, \cdots$.

We show now that if $\sigma(N)$ is $n + 1$ discount optimal for chain N and, with no loss of generality, $\sigma(N) = \sigma(M)$ whenever chains N and M coincide, then $\sigma = (\sigma(N))$ is n discount optimal for the system. To see this suppose $\epsilon > 0$ is given. Then for each N there is a $\rho_N > 0$ with $\rho^{-n-1}(V^{\rho N}_{\sigma(N)} - V^{\rho N}_{\pi(N)}) \geqq -\epsilon$ for all $0 < \rho < \rho_N$ and all $\pi(N)$. In the foregoing we can, without loss of generality, assume $\rho_N = \rho_M$ if chains N and M coincide. Then $\rho^* = \min_N \rho_N > 0$ since there are only finitely many distinct ρ_N. Also for $0 < \rho < \rho^*$

$$\rho^{-n}(\tilde{V}^{\rho}_{\sigma} - \tilde{V}^{\rho}_{\pi}) = \rho \sum_{N=1}^{\infty} \beta^N p_N \rho^{-n-1}(V^{\rho N}_{\sigma(N)} - V^{\rho N}_{\pi(N)}) \geqq -\rho \sum_{N=1}^{\infty} \beta^N \epsilon$$

$$= -\epsilon$$

for all $\pi = (\pi(N))$ so σ is indeed n discount optimal for the system.

It might be hoped that our hypothesis that $\sigma(N)$ is $n + 1$ discount optimal for each chain N would imply the stronger conclusion that $\sigma = (\sigma(N))$ is $n + 1$ discount optimal for the system. That this is not so in general may be seen by specializing to the case where all chains are identical and $p_N = p$ is positive for all N. In that event write $\tilde{V}^{\rho}_{\delta} \equiv \tilde{V}^{\rho}_{\pi}$ where $\pi = (\delta^{\infty}, \delta^{\infty}, \cdots)$. Then from (32) we see that $\rho^{-n}\tilde{V}^{\rho}_{\delta} = p\rho^{-n-1}V^{\rho}_{\delta}$. Thus $\pi = (\delta^{\infty})$ is n discount optimal for the system if and only if δ^{∞} is $n + 1$ discount optimal for each of the (identical) individual chains.

As a specific illustration of this last situation, suppose each chain is like that in Figure 2, and let $\delta^1 = \alpha$ and $\gamma^1 = \beta$. Observe that if you always start in state one and use δ^∞ for each chain, then at time one you earn one thousand dollars. And at each subsequent time $N > 1$ your net earnings are zero, resulting from a one thousand dollar income in chain N and a one thousand dollar expense in chain $N - 1$. The upshot is that your total system earnings under δ^∞ starting from state one is one thousand dollars. This is so in spite of the fact that each individual chain earns zero dollars under δ^∞! The one thousand dollar permanent system earnings results entirely from aggregating the temporary earnings of each chain during complementary periods of time. In this example, γ^∞ is 0 discount optimal for each chain but only δ^∞ is 0 discount optimal for the system.

BIBLIOGRAPHIC NOTES

The main references for results in this paper are given below, though attribution of some intermediate results is omitted. For related results on Markov decision chains, see the books by Howard [12] and Derman [7], and Denardo's paper [6].

Section 2. The optimality of bold and timid play in example one was established respectively by Dubins and Savage [10] and Ross [17]. General conditions assuring a myopic stopping rule is optimal in example two were first given by Derman and Sacks [8] and Chow and Robbins [3], though our treatment follows Breiman [2]. The treasure hunting problem is due to Ronald A. Howard (unpublished) and is included here with his kind permission. Scarf [18] first showed an (s', S') policy is optimal in example three when $H(\cdot)$ is convex. The weaker hypothesis used here is due to Veinott [21]. Johnson [13] suggested how the policy improvement method could be used to find an optimal (s', S') policy efficiently. Each of the three examples is merely representative of an entire area of study to which the reader will be referred by consulting [10], [4], [23], [20].

Section 3. Shapley [19] introduced Markov decision chains and proved Lemma 1 by successive approximations under the hypothesis that the row sums of the transition matrices are all less than one. Under this same hypothesis, Howard [12] proved (8), devised the policy improvement method, and used it to give a constructive proof of Lemma 2; and Blackwell [1] proved Theorem 2. The final form of Lemma 2 was established by Denardo [5] and that of the remaining results by Veinott [22]. I am grateful to Eric V. Denardo (private communication, May 1969) for the new constructive proof that 1° implies 3° of Theorem 1.

Section 4. The proof of (15) is borrowed from Doob [9]. Kemeny and Snell [15] showed $P^* - Q$ is nonsingular where P is irreducible or transient, and Blackwell [1] observed the result holds in general. The proof given here is due to Veinott [22]. The Laurent expansion in Theorem 3 appears in Hille and Philips [11] (in an abstract continuous time version). It was introduced and proved by matrix methods in Markov decision chain theory by Miller and Veinott [16]. The resolvent approach used here is taken from Veinott [22] which in turn borrowed from Kato [14]. Blackwell [1] introduced the notion of discount optimality and proved simply, but nonconstructively, the existence of a stationary discount optimal policy. The remaining results of the section are due to Miller and Veinott [16]. Theorem 5 incorporates a refinement suggested by Denardo and other improvements.

Section 5. The definition and characterization of n discount optimality was introduced by Blackwell [1] for $n = 0$ and by Veinott [22] for $n \neq 0$. Howard [12] devised a policy improvement method for finding a -1 discount optimal stationary policy and Blackwell [1] established the variant used here for that case. The remaining results are due to Veinott [22], though the proof of Theorem 6 is new. The interpretation of 1 discount optimality and the treatment of systems of Markov decision chains are also new.

REFERENCES

1. Blackwell, D., "Discrete dynamic programming," *Ann. Math. Statist.*, **33** (1962), 719–726.
2. Breiman, L., "Stopping-rule problems," in E. F. Beckenbach (ed.), *Applied Combinatorial Mathematics*, New York: Wiley, 1964, pp. 284–319.

3. Chow, Y. S., and H. Robbins, "A martingale system theorem and applications," *Proc. Fourth Berkeley Symposium Math. Statist. Prob.*, 1, Berkeley: Univ. Calif. Press, 1961, pp. 93–104.

4. ———, and D. Siegmund, *Great Expectations: The Theory of Optimal Stopping*, Boston: Houghton Mifflin, 1971.

5. Denardo, E. V., "Contraction mappings in the theory underlying dynamic programming," *SIAM Review*, 9 (1967), 165–177.

6. ———, "A Markov decision problem," in T. C. Hu and S. M. Robinson (eds.), *Mathematical Programming*, New York: Academic Press, 1973.

7. Derman, C., *Finite State Markovian Decision Processes*, New York: Academic Press, 1970.

8. ———, and J. Sacks, "Replacement of periodically inspected equipment (an optimal stopping rule)," *NRLQ*, 7 (1960), 597–607.

9. Doob, J. L., *Stochastic Processes*, New York: Wiley, 1953, pp. 175–176.

10. Dubins, L. E. and L. J. Savage, *How to Gamble if You Must: Inequalities for Stochastic Processes*, New York: McGraw Hill, 1965, pp. 87–89.

11. Hille, E., and R. S. Philips, *Functional Analysis and Semigroups* (rev. ed.), American Mathematical Society, Providence, R.I., 1957.

12. Howard, R. A., *Dynamic Programming and Markov Processes*, New York: Wiley, 1960.

13. Johnson, E. L., "On (s, S) policies," *Man. Sci.*, 15 (1968), 80–101.

14. Kato, T., *Perturbation Theory for Linear Operators*, New York: Springer-Verlag, 1966, pp. 38–40.

15. Kemeny, J. G., and J. L. Snell, *Finite Markov Chains*, Princeton: Van Nostrand, 1960, pp. 100–101.

16. Miller, B. L., and A. F. Veinott, Jr., "Discrete dynamic programming with a small interest rate," *Ann. Math. Statist.*, 40 (1969), 366–370.

17. Ross, S., "Dynamic programming and gambling models," ORC 72-24, Operations Research Center, Berkeley: University of California, 1972.

18. Scarf, H., "The optimality of (S, s) policies in the dynamic inventory problem," Chapter 13 in K. Arrow, S. Karlin, and P. Suppes (eds.), *Mathematical Methods in the Social Sciences*, Stanford, Calif.: Stanford Univ. Press, 1960.

19. Shapley, L. S., "Stochastic games," *Proc. Nat. Acad. Sci. U.S.A.*, 39 (1953), 1095–1100.

20. Veinott, Jr., A. F., "The status of mathematical inventory theory," *Man. Sci.*, 12 (1966), 745–777.

21. ———, "On the optimality of (s, S) policies: New conditions and a new proof," *SIAM J. Appl. Math.*, 14 (1966), 1067–1083.

22. ———, "Discrete dynamic programming with sensitive discount optimality criteria," *Ann. Math. Statist.*, 40 (1969), 1635–1660.

23. Wald, A., *Sequential Analysis*, New York: Wiley, 1947.

THE DECOMPOSITION ALGORITHM FOR LINEAR PROGRAMS

George B. Dantzig
and
Philip Wolfe

0. ABSTRACT

A procedure is presented for the efficient computational solution of linear programs having a certain structural property characteristic of a large class of problems of practical interest. The property makes possible the decomposition of the problem into a sequence of small linear programs whose iterated solutions solve the given problem through a generalization of the simplex method for linear programming.

1. THE DECOMPOSED LINEAR PROGRAM

Many linear programming problems of practical interest have the property that they may be described, in part, as composed of separate linear programming problems tied together by a number of constraints considerably smaller than the total number imposed on the problem. When the matrix of coefficients of such a problem, suitably ordered, is displayed in the usual way, a pattern emerges like that shown in Figure 1. In this figure the constraint matrix has been partitioned into nonzero blocks A_j and B_j, the right-hand side column of constants correspondingly into b, b_1, $\cdots$, b_n; and the "costs," the coefficients of the objective form, into c_1, c_2, $\cdots$, c_n.

More particularly, suppose that we have:

an m-vector b, and, for each $j = 1$, $\cdots$, n,
A_j, an m by n_j matrix,
B_j, an m_j by n_j matrix,
c_j, an n_j-vector, and
x_j, a variable n_j-vector.

Then the problem illustrated in Figure 1 is posed as a linear programming problem in $\Sigma_j\, n_j$ variables, subject to $m + \Sigma_j\, m_j$ constraints:

The Decomposed Program. Find the vectors x_j $(j = 1, \cdots, n)$

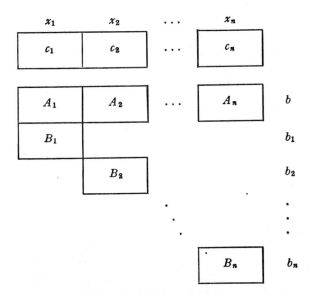

FIG. 1. The Original Problem.

satisfying the constraints

(1) $$\sum_j A_j x_j = b \qquad \text{and} \qquad x_j \geqslant 0 \text{ (all } j\text{)}$$

and

(2) $$B_j x_j = b_j \text{ (all } j\text{)}$$

which minimize the linear form

(3) $$\sum_j c_j x_j.$$

It would seem that each of the n sets of constraints of (2) constitutes a "subproblem" of secondary importance to the whole program, and that they should be studied mainly through the restrictions they impose on the activities of the "joint" constraints (1). Pursuing this point of view leads us from the decomposed program to the equivalent *extremal problem* of the next section. For the sake of simplicity in formulating the extremal problem, it is assumed that

(4) $$S_j = \{x_j \mid x_j \geqslant 0, \qquad B_j x_j = b_j\}$$

is bounded for each j. Removing this restriction in the treatment which follows requires only minor changes, which are discussed in Section 4.

The results of this paper can be viewed as arising from the statement of the decomposed program as a "generalized linear programming problem" in n variables, in which the m-element column of coefficients associated with each variable is drawn freely from the convex set S_j instead of being fixed as in the ordinary linear programming problem. This generalized problem will be developed more fully elsewhere.

2. THE EXTREMAL PROBLEM

This section will formulate a type of linear programming problem in which the features of the decomposed program (1)–(3) pertaining to its joint constraints are brought to the fore. In the new problem, the extreme points of the sets defined by the relations (2) will yield the data for the problem.

For $j = 1, \cdots, n$, let $W_j = \{x_{j1}, \cdots, x_{jK_j}\}$ be the set of all extreme points of the convex polyhedron S_j defined by the conditions $x_j \geqslant 0$, $B_j x_j = b_j$; also define

$$
(5) \qquad \left.\begin{array}{c} P_{jk} = A_j x_{jk} \\[2mm] c_{jk} = c_j x_{jk} \end{array}\right\} \text{for } k = 1, \cdots, K_j.
$$

The Extremal Program. Find numbers s_{jk} ($j = 1, \cdots, n$; $k = 1, \cdots, K_j$) satisfying

$$
(6) \qquad \sum_{j,k} P_{jk} s_{jk} = b, \; s_{jk} \geqslant 0 \qquad (\text{all } j, k),
$$

and

$$
(7) \qquad \sum_k s_{jk} = 1 \qquad (\text{all } j)
$$

which minimize the linear form

$$
(8) \qquad \sum_{j,k} c_{jk} s_{jk}.
$$

The relation of the extremal problem to the original problem lies in the fact that any point x_j of S_j, because it is bounded and a convex polyhedral set, may be written as a convex combination of its extreme points, that is, as $\Sigma_k x_{jk} s_{jk}$, where $\{s_{j1}, \cdots, s_{jK_j}\}$ satisfy (7); and the expressions (6) and (8) are just the expressions (1) and (3) of the decomposed problem rewritten in terms of the s_{jk}. This relation is stated in the following lemma, which needs no proof.

LEMMA: *If the numbers $\{s_{jk}\}$ solve the extremal program* (6)–(8), *then the vectors*

$$
(9) \qquad s_j = \sum_k x_{jk} s_{jk} \qquad (j = 1, \cdots, n)
$$

solve the problem (1)–(3).

Note that by (5) and (9) $\Sigma_j A_j x_j = \Sigma_j \Sigma_k A_j x_{jk} s_{jk} = b$ and $\Sigma_j c_j x_j = \Sigma_{jk} c_{jk} s_{jk}$.

The matrix of coefficients for the extremal problem is displayed in Figure 2. Its constraint equations are $m + n$ in number; the m joint constraints of the original problem have gone over into the m constraints (6) constituting the upper block in Figure 2, and the m_j constraints of the jth subproblem have gone over into single constraints of the form (7). The reduction in the total number of constraints is sizeable in case the m_j are large, and it is this fact on which the computational efficiency of the decomposition principle relies. The reduction appears to have been accomplished, however, by greatly enlarging the number of variables in the problem from the original $\Sigma_j\, n_j$ to the number $\Sigma_j\, K_j$; the proposed method would be of little interest or value if it were not possible effectively to reduce this number.

3. THE DECOMPOSITION ALGORITHM

The central idea of the decomposition principle is that the extremal linear programming problem of (6)–(8) and Figure 2 can be solved by the simplex method for linear programming without

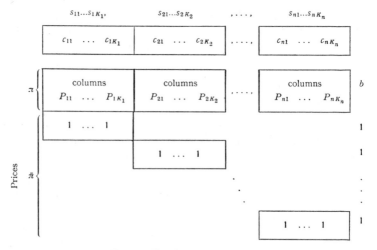

FIG. 2. The Extremal Problem

prior calculation of all the data given in the statement of the problem. Typically, only a small number of the $\Sigma_j K_j$ activities present will ever play an active role in the course of solving it, and, as will be seen, the coefficients needed in handling just these activities may be generated at the time they are to come under consideration. The principle of coefficient generation was first employed in the study of multi-commodity network flows [4].

Let us repeat here the essential features of the simplex method that will be used. The phenomenon of degeneracy plays the same role in the extremal problem that it does in any linear programming problem, and standard methods [1] can be invoked to handle it when necessary, so that it need not be discussed here. The problem of finding a first feasible solution to the extremal problem is similar to the same problem for a general linear program and is discussed in the next section. It will now be supposed that the usual data are at hand for performing one typical step of the simplex method on the extremal problem.

Since the extremal problem has $m + n$ equation constraints, there will be at hand $m + n$ columns ($m + n$-component vectors corresponding to extreme points) which constitute a "feasible basis;" that is, these columns are linearly independent, and the unique solution of the constraint equations ((6) and (7)) obtained by setting to zero those variables associated with all other columns is nonnegative. If the revised simplex method is used in performing the calculations, there will also be at hand the $m + n$-vector of "prices" (π; $\bar{\pi}$)—the m-vector π being associated with the first m constraints and the n-vector $\bar{\pi}$ with the remaining n, as in Figure 2. In the general linear program, the inner product of the price vector with any column of the basis must equal the cost associated with that column; in the case of the extremal problem, letting $\bar{\pi}_j$ be the jth component of $\bar{\pi}$, this relationship can be written

$$(10) \qquad \pi P_{jk} + \bar{\pi}_j = c_{jk}$$

for basic columns drawn from the jth partition, $j = 1, \cdots, n$.

One step of the simplex method iteration for solving the extremal problem would be performed as follows: find a column of the constraint matrix whose "reduced cost" is negative, that is, for which

$$(11) \qquad c_{jk} - \pi P_{jk} - \bar{\pi}_j < 0$$

(commonly the column which minimizes the reduced cost is chosen); add this column to the current basis, and delete one column from the basis in such a way that the new basis is still feasible. If no column satisfying (11) can be found, then the current solution $\{s_{jk}\}$ solves the extremal problem. Otherwise, the simplex method gives the appropriate rules for the removal of a column from the basis and for the calculation of the new prices $(\pi; \bar{\pi})$ associated with the new basis, with which the next iterative step can begin.

A procedure for applying the simplex algorithm to the extremal problem without having all the data of that problem at hand can now be stated. It is supposed that only the $m + n$ columns of a feasible basis for the problem are given.

The Decomposition Algorithm—Iterative Step. Given $m + n$ columns of the form $(P_{jk}; 0, \cdots, 1, \cdots, 0)$ which constitute a feasible basis for the extremal problem, given their associated costs c_{jk}, and given prices $(\pi; \bar{\pi})$ satisfying (10), then for each $j = 1, \cdots, n$ let $\bar{x}_j$ be an extreme point of S_j minimizing the linear form

$$(12) \qquad (c_j - \pi A_j)x_j,$$

under the constraints

$$x_j \geqslant 0, \qquad B_j x_j = b_j;$$

and let $\bar{x}_{j_0}$ be such that

$$(13) \qquad \delta = (c_{j_0} - \pi A_{j_0})\bar{x}_{j_0} - \bar{\pi}_{j_0} = \mathrm{Min}_j[(c_j - \pi A_j)\bar{x}_j - \bar{\pi}_j].$$

If $\delta < 0$, form the new column and its associated cost for the extremal problem as

$$(14) \qquad (A_{j_0}\bar{x}_{j_0}; 0, \cdots, 1, \cdots, 0) \quad \text{and} \quad c_{j_0}\bar{x}_{j_0}.$$

Add this column to the basis and form a new basis and new prices using the rules of the simplex method. If $\delta \geqslant 0$, terminate the algorithm; $\{s_{jk}\}$ solves the extremal problem, and the relations (9) give the solution of the original problem.

THEOREM: *The decomposition algorithm terminates in a finite number of iterations, yielding a solution of the extremal problem.*

Proof: Since the termination of the simplex method as applied to the general linear programming problem in a finite number of iterations is known [1], it is sufficient to show that the rules of the step of the decomposition algorithm yield a column satisfying the criterion (11), if this is possible. For the column defined by (14), the left-hand side of (11) is just $c_{j_0}\bar{x}_{j_0} - \pi A_{j_0} - \pi_{j_0} = \delta$, which is as small as possible. (The lemma of Section 2 shows that a solution of the original problem is obtained when a solution of the extremal problem has been found by this algorithm.)

In summary, a cycle of the decomposition algorithm can be stated this way:

Given $m + n$ columns constituting a feasible basis for the extremal problem, use the prices associated with that basis to form the *modified costs* $c_j - \pi A_j$ for each of the *subproblems*:

$$\text{Minimize } (c_j - \pi A_j)x_j \quad \text{for} \quad x_j \geqslant 0, \qquad B_j x_j = b_j.$$

If $\delta < 0$, so that the original problem is not solved, then from an appropriate subproblem solution construct a new column for the extremal problem and form a new feasible basis incorporating that column, deleting another column from the basis in accordance with the rules of the simplex method.

4. DETAILS ON THE ALGORITHM

This section considers some details of the use of the decomposition algorithm: getting the algorithm started, dealing with unbounded solutions of the subproblems when the boundedness restriction is removed, and some variations of the selection technique (12) and of methods of decomposing a problem.

Initiating the Algorithm. The decomposition algorithm can be started with precisely the same device, called Phase One, used for the general linear programming problem. This device consists in augmenting the problem with $m + n$ "artificial" variables in terms of which an initial basic feasible solution, and the prices associated with the corresponding initial feasible basis, are readily found. The decomposition algorithm can then be applied to the problem of removing the artificial variables. After this has been done, the required starting conditions for the ordinary application of the algorithm are automatically met.

For $i = 1, \cdots, m + n$: let y_i be a nonnegative variable; let I_i be the ith column of the $m + n$-order identity matrix; and let $c_i = 1$ be the cost associated with the variable y_i. For Phase One, replace all the costs $\{c_j\}$ of the original problem with zero vectors. It is assumed without loss of generality that the right-hand side vectors b, $\{b_j\}$ are all nonnegative.

Designating $\{I_i\}$ as the initial feasible basis, employ the decomposition algorithm in the minimization of the linear form $\Sigma_i \, y_i$. (Note that the initial feasible solution and the initial prices are given by $(y_1, \cdots, y_{m+n}) = (b; b_1, \cdots, b_n)$ and $(\pi; \bar{\pi}) = (1, \cdots, 1)$.)

If the form $\Sigma_i \, y_i$ cannot be reduced to zero, then the extremal problem has no feasible solution; if the form can be reduced to zero, then a feasible solution is a fortiori at hand. Typically, this Phase One will have ended with none of the artificial columns left in the basis. (It occasionally happens that an artificial column remains in the basis although its variable is, of course, zero. Handling this case requires the use of an additional constraint in the problem, described elsewhere [1].) The cost vectors of the original problem are then restored, and Phase Two, the application of the algorithm to the proper extremal problem, can proceed as in Section 3.

Extension to the Unbounded Case. It was assumed in (4) that the constraint set S_j of each subproblem was bounded. If this is not the case, then it may happen in applying the decomposition algorithm that an unbounded solution is obtained for the subproblem

(15)   Minimize $(c_j - \pi A_j)x_j$ under the constraints

$$x_j \geqslant 0, \qquad B_j x_j = b_j.$$

The only change incurred in the algorithm by this is that the rule for forming a new column for the extremal problem must be extended. An unbounded solution of the problem (15) will cause the simplex method to yield a vector y_j such that

(16) $y_j \geqslant 0$, $B_j y_j = 0$,   and $(c_j - \pi A_j)y_j < 0$,

the direction of an infinite ray of feasible solutions along which the cost form of (15) proceeds to $-\infty$.

In this case, the column and cost to be added to the extremal problem have the form

(17) $(A_j y_j; 0, \cdots, 0)$ and $c_j y_j$

instead of the form (14); the "1" has been omitted from the column so that the variable s_{jk} associated with this column will not be constrained by the relation $\Sigma_k \, s_{jk} = 1$ imposed upon the columns created from bounded solutions; any nonnegative multiple of this column is admissible, corresponding to the fact that any nonnegative multiple of the vector y_j is admissible in the representation of the constraint set of (15) as the sum of a bounded polyhedron and a polyhedral convex cone. It is easy to check that with this extra freedom the decomposition algorithm still consists only in the application of the simplex method to the extremal problem thus extended, and hence has the same termination properties as in the bounded case since the number of possible rays generated by the simplex solutions of the subproblems is finite.

Variations of the Decomposition Process. It is possible to pursue the decomposition algorithm in different ways for the sake of computational efficiency. One suggestion is easily made: retain as many of the solutions $\bar{x}_j$ to the subproblems (12) as would provide new basic columns for the extremal problem; that is, for each j such that $(c_j - \pi A_j)\bar{x}_j < \bar{\pi}_j$, form the column $(A_j \bar{x}_j; 0, \cdots, 1, \cdots, 0)$ and its cost $c_j \bar{x}_j$, and adjoin all these to the given basic columns, employing the simplex method to find a minimum among this extended set of columns. This procedure, while requiring perhaps more simplex method iterations, should require fewer subprogram solutions to be found.

It should be noted that finding the n subproblem solutions of (12) at each iteration of this procedure may not require the complete solution of n linear programs. If the prices in the extremal problem do not change much from one iteration to the next, then the new subproblem solutions may not differ much, if at all, from their old, in which case the new solutions can be found with little labor. Also, in many practical cases, the subproblems themselves have a structure such that their solutions are readily found by means of special devices, for example, when the subproblems are transportation problems.

Another variation in the use of decomposition consists in the fact that a given linear programming problem may be decomposed to various degrees. For example, in the original problem (1)–(3) pictured in Figure 1, the first two partitions might have been considered collectively as one, embracing the single vector $(x_1; x_2)$. Little would be changed in the description of the decomposition algorithm, except that the first subproblem of the family (12) would read

(18)   Minimize $(c_1 - \pi A_1)x_1 + (c_2 - \pi A_2)x_2$ under the restrictions

$$(x_1, x_2) \geqq 0, \qquad B_1 x_1 = b_1, \qquad B_2 x_2 = b_2,$$

and the column constructed therefrom would be

(19) $(A_1 \bar{x}_1 + A_2 \bar{x}_2; 0, \cdots, 1, \cdots, 0)$,   where $(\bar{x}_1, \bar{x}_2)$ solves (18).

The two variables of the problem (18), however, are completely independent, so that the solution $(\bar{x}_1; \bar{x}_2)$ is just composed of the solutions $\bar{x}_1$, $\bar{x}_2$, of the two separate problems of the ordinary form for $j = 1, 2$. The result, then, of "aggregating" the first two partitions has been to aggregate their resultant columns according to the formula (19).

If this aggregation is carried as far as possible, all the n subproblems may be taken together as one. On account of the independence of the n parts of this one subproblem, finding the subproblem solution involves the same work as before. An advantage, however, derives from the fact that the extremal problem now has only $m + 1$ constraints, instead of $m + n$, and thus is easier to handle. The decomposition of a linear programming problem into a problem having only one partition, when applied to the classical transportation problem, yields an algorithm which promises considerable efficiency in the case of transportation from a small number of origins to a large number of destinations [5].

Finally, it should be mentioned that the linearity of the cost functions $c_j x_j$ for the partitions of the problem is not essential for the application of the decomposition principle. Suppose instead that in each partition this function were replaced by the function $f_j(x_j)$, which, for success of the method, need only be convex. The decomposition algorithm would be unchanged, except that the subproblems (12) would assume the form

(20)   Minimize $f_j(x_j) - \pi A_j x_j$ under the constraints

$$x_j \geqslant 0, \qquad B_j x_j = b_j.$$

Solutions $\bar{x}_j$ of these problems would then be used to construct new extremal columns by formula (14) with associated costs $f_j(\bar{x}_j)$. In the nonlinear case, termination of the decomposition algorithm is not assured, but convergence to a solution of the problem is. The arguments used differ from those above; this extension of the decomposition algorithm is given elsewhere [2].

5. A NUMERICAL EXAMPLE

The small problem we shall solve by means of the decomposition algorithm is somewhat artificial, but will show the utility of the method in even the case of a single partition.

A homogeneous commodity is to be transported from each of two sources to each of four destinations. The costs of transportation of one unit of the commodity from a given source to a given destination appears as an entry in the matrix of Figure 3. Beside the matrix and below it are written respectively the amounts of the commodity available from each source and demanded at each destination. One requirement is added which prevents this from being an ordinary transportation problem: it is required that three times the quantity shipped from source 1 to destination 3 and twice the quantity shipped from source 2 to destination 2 sum to 10.

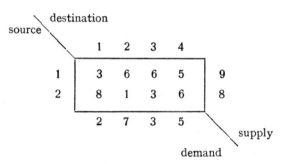

FIG. 3. Data for the Example

In order to formulate this as a linear programming problem, denote by t_{ij} ($i = 1, 2; j = 1, \cdots, 4$) the quantity to be shipped from source i to destination j. The coefficients of all the constraints are shown in Figure 4. Since there is only one partition, subscripts are not needed to describe the parts of the matrix. The row constituting the matrix A represents the joint constraint on t_{13} and t_{22}. The block B consists of six equations asserting that the net flow from each source must be just the amount available, and that the net flow to each destination must be what is demanded.

A has only one row, and the problem has only one partition, so that $m = 1$ and $n = 1$. The extremal problem is thus a two-equa-

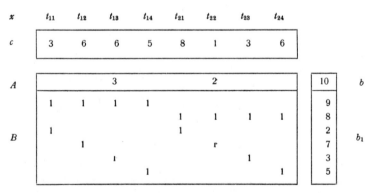

FIG. 4. Decomposition of the Example

3	6	$6-3\pi$	5	9
8	$1-2\pi$	3	6	8
2	7	3	5	

FIG. 5. The Subproblem as a Transportation Problem

tion problem, and the price vector $(\pi; \bar{\pi})$ has two components. Of course, $\{t_{ij}\} = x$. The single subproblem of (12), the problem of minimizing $(c - \pi A)x$ under the constraints $Bx = b$, becomes here just that of solving the transportation problem shown in Figure 5, since the subproblem constraints are precisely those of the transportation problem of Figure 3.

Rapid special methods for solving transportation problems are well known [3]; we need not go into them here.

DECOMPOSITION ALGORITHM ITERATIONS FOR THE EXAMPLE

Iteration no., k	Basic solution	π	$\bar{\pi}$	Subproblem solution, x	Extremal cost	Extremal column, P_k	δ
1	$10\,I_1 + I_2$	1	1	$\begin{bmatrix}1 & & 3 & 5\\ 1 & 7 & & \end{bmatrix}$	0	$\begin{bmatrix}23\\1\end{bmatrix}$	–24
2	$\frac{10}{23}P_1 + \frac{13}{23}I_2$	$-\frac{1}{23}$	1	$\begin{bmatrix}2 & 7 & & \\ & & 3 & 5\end{bmatrix}$	0	$\begin{bmatrix}0\\1\end{bmatrix}$	–1
3	$\frac{10}{23}P_1 + \frac{13}{23}P_2$	0	0	End of Phase One			
[Phase Two begins with costs for P_1, P_2: 61, 87.]							
3 Cont.		$-\frac{26}{23}$	87	$\begin{bmatrix}2 & & 2 & 5\\ & 7 & 1 & \end{bmatrix}$	53	$\begin{bmatrix}20\\1\end{bmatrix}$	$-11\frac{9}{23}$
4	$\frac{1}{2}P_3 + \frac{1}{2}P_2$	$-\frac{17}{10}$	87	$\begin{bmatrix}2 & 2 & & 5\\ & 5 & 3 & \end{bmatrix}$	57	$\begin{bmatrix}10\\1\end{bmatrix}$	–13
5	$0\,P_3 + 1\,P_4$	$-\frac{2}{5}$	61	$\begin{bmatrix}2 & 2 & & 5\\ & 5 & 3 & \end{bmatrix}$			0

Since $\delta \geqslant 0$, the basic solution is optimal. The objective function has the values $75\frac{16}{23}$, 70, 57 in the three stages of Phase Two.

Beginning with a basic feasible solution for the extremal problem constructed from the artificial columns I_1, I_2 (see Section 4), the major data generated in the decomposition algorithm solution of this problem are given in the table on page 173. We have omitted only the inverse of each feasible basis, which is generally carried along in simplex method calculations.

The RAND Corporation

REFERENCES

1. Dantzig, G. B., A Orden, and P. Wolfe, "The generalized simplex method for minimizing a linear form under linear inequality constraints," *Pacific Journal of Mathematics*, **5** (1955), 183–195.
2. Dantzig, G. B., "General convex objective functions," *Mathematical Methods in the Social Sciences*, editors, Arrow, Karlin, and Suppes, Chapter 10, Stanford University Press, June 1960.
3. Ford, L. R., Jr., and D. R. Fulkerson, "Solving the transportation problem," *Management Science*, **3** (1956), 24–32.
4. ———, "A suggested computation for maximal multi-commodity network flows," *Management Science*, **5** (1958), 97–101.
5. Williams, A. C., "A treatment of transportation problems by decomposition", *SIAM J.*, **10** (1962), 35–48.

INDEX